A REFERENCE CATALOGUE AND ATLAS OF GALACTIC NOVAE

A REFERENCE CATALOGUE AND ATLAS OF GALACTIC NOVAE

by

HILMAR W. DUERBECK
University of Munster, F.R.G.

Reprinted from Space Science Reviews, Vol. 45, Nos. 1–2

D. Reidel Publishing Company
Dordrecht / Boston

Library of Congress Cataloging-in-Publication Data

CIP

Duerbeck, Hilmar W., 1948–
A reference catalogue and atlas of galactic novae.

Bibliography: p.
Includes index.
1. Stars, New – Atlases. 2. Stars, New – Catalogs.
3. Milky Way – Atlases. 4. Milky Way – Catalogs.
I. Title.
QB841.D84 1987 523.8′446 87–12688
ISBN 90–277–2535–7

Published by D. Reidel Publishing Company,
P.O. Box 17, 3300 AA Dordrecht, Holland.

Sold and distributed in the U.S.A. and Canada
by Kluwer Academic Publishers,
101 Philip Drive, Norwell, MA 02061, U.S.A.

In all other countries, sold and distributed
by Kluwer Academic Publishers Group,
P.O. Box 322, 3300 AH Dordrecht, Holland.

Printed in Belgium

A REFERENCE CATALOGUE AND ATLAS OF GALACTIC NOVAE*

HILMAR W. DUERBECK

Astronomisches Institut der Universität Münster, Wilhelm-Klemm-Str. 10, D-4400 Münster, F.R.G.

(Received 31 December, 1986)

Idem Hipparchus ... novam stellam in aevo suo genitam deprehendit ... adnumerare posteris stellas ac sidera ad nomen expungere organis excogitatis, per quae singularum loca atque magnitudines signaret, ut facile discerni posset ex eo ... an obirent ac nascerentur ... caelo in hereditate cunctis relicto, si quisquam, qui cretionem eam caperet, inventus esset.

C. Plinius Sec., *Nat. Hist.* 1, II, XXIV (AD 77)

Meditations of evolution increasingly vaster: of moribund and of nascent new stars such as Nova in 1901...

James Joyce, *Ulysses* (1922)

Auch die jetzt allgemein angenommene Erklärung solcher Novae: daß die Bewohner zu viel Atomversuche angestellt hätten, und es sich also um ein ganz natürliches Stadium der Sternentwicklung dabei handele.

Arno Schmidt, *Die Gelehrtenrepublik* (1957)

Abstract. This catalogue and atlas contains information on 277 objects, mainly classical novae and related objects (recurrent novae, X-ray novae, dwarf novae with long cycle lengths, symbiotic stars and suspected new stars). For most objects, brightness ranges, accurate positions, finding charts and selected bibliographies are given.

1. Objectives of the Reference Catalogue and Atlas of Galactic Novae

Some novae are spectacular objects during their outburst phases, yet all novae are new because no light curve or spectral appearance exactly matches that of a predecessor. Novae are objects whose radiation is generated and transformed under unusual, non-stationary conditions: accretion, thermonuclear runaways, expanding gas shells and dust shells formed and destroyed.

Novae are interesting at maximum as well as at minimum light. Astronomers, planning to observe them during maximum light, consult the current IAU Circulars and Telegrams. The present catalogue and atlas of galactic novae intends to serve those who

* Based in part on observations collected at the European Southern Observatory, La Silla, Chile, the Centro Astronomico Hispano-Aleman Calar Alto, operated by the Max-Planck-Institut für Astronomie, Heidelberg, and on measurements made at the European Southern Observatory, Garching, F.R.G.

Space Science Reviews **45** (1987) 1–212.

wish to observe novae at minimum light and require information about their location, magnitude and outburst properties.

This catalogue and atlas is especially designed for the observer. It contains all classical novae, recurrent novae, X-ray novae, and in addition also objects once believed to be novae, but now classified as other types of variables – albeit often related ones: symbiotic stars and dwarf novae. Objects with insufficient evidence, including some which may not exist at all, are listed to encourage further studies towards clarification of the cases. Anyone who uses the catalogue for statistical work should consult the 'classification' entry for each object.

Generally, novae at minimum are faint. Nevertheless, many of them are within the reach of modern photometers, spectrographs, and CCDs. The lack of observations seems to result in part from the lack of accurate positions (or, at least, of easily accessible ones), and of detailed finding charts. Accurate positions are essential for observations in other wavelength regions, especially from satellites.

The present catalogue and atlas contains:

- positions (equinox 1950.0), with accuracies of 1″ or better, for all novae, suspected novae and related objects for which reliable data are published or available from the study of accessible plate collections;
- finding charts, prepared from sky atlas plates;
- apparent magnitudes at maximum and minimum light;
- light curve types according to the classification scheme of Duerbeck (1981) and t_3-times;
- principal bibliographical references relating to light curves, spectra, nebulosities, duplicities, and additional features;
- notes on the classification.

2. Previous Catalogues and Atlases

Table I gives bibliographical data of old catalogues containing novae (general catalogues of variable stars excluded) and more recent catalogues of ancient novae. Generally, they do not furnish sufficiently accurate positions.

Table II lists articles about 'missing stars', predominantly catalogue entries based on visual observations of stars which were later reported missing. Only those objects are included in the present catalogue and atlas whose original classification as novae has not been challenged. The low positional accuracy and poor documentation of the observations makes identifications almost impossible.

Table III lists catalogues of novae. The most complete catalogue was compiled by Payne-Gaposchkin (1957, supplement 1977). It lacks, however, precise positions and finding charts. Khatisov (1971), Wyckoff and Wehinger (1978) and Williams (1983) published accurate positions and finding charts for 42, 25, and 17 novae, respectively. Unfortunately, each of the subsequent writers was apparently unaware of the earlier papers and the same well-known objects were studied over again. The recent publications by Li *et al.* (1984a, b) and Liu and Li (1984) contain positions and finding charts

TABLE I

Catalogues of historical novae (prior to about 1850)

Königl. Preuß. Akad. Wissensch.: 1766, 'Verzeichniss der bisher bemerkten neuen und veränderlichen Sterne', Tafel XV in: *Sammlung astronomischer Tafeln*, I. Band, Berlin, p. 212.
Pingré, A. G.: 1783, *Cométographie ou traité historique et théorique des comètes*, Vol. 1, Paris.
Biot, E.: 1843, Catalogue des Étoiles extraordinaires observées en Chine depuis les temps anciens jusqu'à l'an 1203 de notre ère, *Connaissance des temps pour l'an 1846*, Paris, p. 60.
Chambers, G. F.: 1867, 'A Catalogue of "New Stars"', *Mon. Not. R. Astr. Soc.* **27**, 243.
Zinner, E.: 1919, 'Die Neuen Sterne', *Sirius* **52**, 25.
Lundmark, K.: 1921, 'Suspected New Stars Recorded in Old Chronicles and Among Recent Meridian Observations', *Publ. Astr. Soc. Pacific* **33**, 225.
Shigeru, K.: 1935, *Nihin Temmon Shiryo*, Wōseisha, Tokyo (quoted by F. R. Stephenson (1976)).
Hsi Tsê-tsung: 1955, 'A New Catalogue of Ancient Novae', *Acta Astron. Sinica* **3**, 183 = 1957, *Astron. Zhurn.* **34**, 159 = 1958, *Smithsonian Contr. Astrophys.* **2**, No. 6.
Ho Peng-Yoke: 1962, 'Ancient and Medieval Observations of Comets and Novae in Chinese Sources', *Vistas in Astronomy* **5**, 127.
Xi Ze-Zong, Po Shu-jen: 1965, 'Ancient Oriental Records of Novae and Supernovae', *Acta Astron. Sinica* **13**, 1 = 1966, NASA Techn. Transl. TTF-388 = 1966, *Science* **153**, 597.
Ho Peng-Yoke, Ang Tian-Se: 1970, 'Chinese Astronomical Records on Comets and "Guest Stars"', *Oriens Extremus* **17**, 63.
Pskovsky, Yu. P.: 1972, 'Survey of Stellar Outbursts of the Pretelescopic Era', *Astron. Zhurn.* **49**, 31 = 1972, *Soviet Astron. – A.J.* **16**, 23.
Stephenson, F. R.: 1976, 'A Revised Catalogue of Pre-telescopic Galactic Novae and Supernovae', *Quart. J. R. Astron. Soc.* **17**, 121.

TABLE II

Catalogues of 'missing stars'

Zinner, E.: 1922, 'Die vermißten Sterne der BD', *Astron. Abhandl. = Ergänzungsheft der Astron. Nachr.* **4**, No. 2.
Zinner, E.: 1936: 'Die vermißten Sterne', *Astron. Nachr.* **260**, 17.
Himpel, K.: 1942, 'Zum Problem der vermißten Sterne', *Astron. Nachr.* **272**, 271.
Zinner, E.: 1943, 'Zum Lichtwechsel der vermißten Sterne', *Astron. Nachr.* **273**, 262.
Zinner, E.: 1946, 'Die vermißten Sterne des Hevelius', *Kl. Veröff. Remeis-Sternw. Bamberg* **1**, 36.
Zinner, E.: 1952, 'Zur Erklärung des Lichtwechsels der vermißten Sterne', *Kl. Veröff. Remeis-Sternw. Bamberg* **7**.

of 53 and 37 novae, respectively. For the fainter objects, they are, however, unreliable and carry on the misidentifications made in earlier catalogues.

A detailed catalogue and atlas of symbiotic stars, which also includes some objects with nova-like outbursts, was published by Allen (1984). Catalogues and atlases of dwarf novae were prepared by Vogt and Bateson (1982, southern and equatorial objects) and by Bruch *et al.* (1987, northern objects).

TABLE III

Recent catalogues containing novae

Müller, G. and Hartwig, E.: 1920, *Geschichte und Literatur des Lichtwechsels*, Bd. **2**, Poeschel und Trepte, Leipzig, p. 415, 468 (32 objects); 1922, *Geschichte und Literatur des Lichtwechsels*, Bd. **3**, Poeschel und Trepte, Leipzig, p. 98 (15 objects).
Stratton, F. J. M.: 1928, in *Handbuch der Astrophysik*, Bd. **6**, J. Springer, Berlin, p. 254 (70 objects); 1936, in *Handbuch der Astrophysik*, Bd. **7**, J. Springer, Berlin, p. 671 (15 objects).
Payne-Gaposchkin, C. H. and Gaposchkin, S.: 1938, *Variable Stars*, Harvard Observatory, Cambridge, Mass., p. 232 (90 objects).
Tuchenhagen, S.: 1938, *Die Neuen Sterne*, Dissertation, Friedrich-Wilhelms-Universität, Berlin, p. 6 (94 + 5 objects).
Cecchini, G. and Gratton, L.: 1941, *Le Stelle Nove*, Atti R. Accad. Ital., Cl. Sci. Fis. Mat. Nat. **13** = Pubbl. R. Oss. Astron. Milano-Merate, p. 10 (110 objects).
Woronzow-Weljaminov, B. A.: 1953, *Gasnebel und Neue Sterne*, Kultur und Fortschritt, Berlin, p. 706 (102 objects).
Payne-Gaposchkin, C.: 1957, *The Galactic Novae*, North-Holland, Amsterdam, p. 2 (159 objects).
Payne-Gaposchkin, C.: 1958, in *Handbuch der Physik*, Bd. **51**, J. Springer, Berlin, p. 752 (80 objects).
Payne-Gaposchkin, C.: 1977, in M. Friedjung (ed.), *Novae and Related Stars*, D. Reidel, Dordrecht, p. 1 (47 objects).
Ritter, H.: 1984, 'Catalogue of Cataclysmic Binaries, Low-Mass X-ray Binaries and Related Objects', *Astron. Astrophys. Suppl.* **57**, 385 (12 objects).

3. Present Work

Information – primarily on published positions and finding charts – was compiled from the 'Geschichte und Literatur des Lichtwechsels' (Müller and Hartwig, 1918) and from its modern sequel, 'Bibliography of Variable Stars' (Wenzel and Huth, 1981; see also Wenzel, 1981). The 'Astronomischer Jahresbericht', 'Astronomy and Astrophysics Abstracts', and the CDS data bank in Strasbourg were searched for information on published positions, finding charts, light curves, and spectroscopic work. Several years of study in various astronomical libraries permitted the tracing of all the quoted literature.

Several observing runs with the Calar Alto 2.2 m telescope furnished spectroscopic information about northern objects; CCD frames taken in search for nebulosities led to new information about nova ejecta (Seitter and Duerbeck, 1986). A similar survey in the southern hemisphere is in progress, starting with observations made at La Silla in 1979 (Duerbeck and Seitter, 1979).

Most of the new nova identifications were made by comparing modern sky atlas fields with outburst plates – the majority of them from the Harvard College Observatory plate collection. Additional information was obtained at the Landessternwarte Heidelberg, and was provided from the Maria Mitchell Observatory, and other observatories listed in the acknowledgements. The intercomparison of modern sky atlas plates (ESO Quick Blue Survey, ESO/SRC Atlas (red/blue), UK I/SR Atlas, UK Equatorial Survey, Palomar Observatory Sky Survey) led to the identification of more recent novae. A few observations for identification purposes were made with the GPO astrograph at La Silla.

4. Census of the Catalogue

The catalogue contains information on 277 objects.

137 (50%) of them are stars whose outburst spectra or unambiguous minimum characteristics classify them as novae beyond doubt. 123 of them could be identified at minimum.

78 (28%) are stars with amplitudes and light curve forms which makes nova classification likely. 60 of them could be identified at minimum.

Thus, 78% of the objects in this catalogue are confirmed and suspected classical novae.

16 (6%) have properties compatible with both novae and related objects. 15 are identified at minimum.

Furthermore, the catalogue and atlas contains data on 12 (4%) dwarf novae of long cycle length or suspected dwarf novae for which only one outburst has been observed. Examples are WZ Sge and CI Gem.

6 (2%) recurrent novae or suspected recurrent novae are listed. Examples are T CrB and AS Psc.

6 (2%) X-ray novae are listed. Examples are V616 Mon and KY TrA.

6 (2%) symbiotic stars and symbiotic novae are listed. Examples are RT Ser and V352 Aql.

11 (4%) Mira stars or suspected Mira stars, which at some time were believed to be novae, are listed. Examples are V607 Aql and V927 Sgr.

3 (1%) other types of variables or unique variables are listed; η Car, FU Ori, V605 Aql, which might be related to novae. Only one object has been omitted: its large brightness makes a finding chart and one more determinations of its position superfluous, its bibliography is excessive – but otherwise it qualifies as a member of this group: P Cyg.

2 (1%) are non-existing stars. One object originated from the confusion with a minor planet, the other is a second official designation given to an already known nova.

5. Notes for the Use of the Catalogue and Atlas

Entries are made in the following order:

Designations and types of variability

The designation of the object is in most cases that of the 'General Catalogue of Variable Stars' (Kholopov, 1985, and earlier editions). When it is not listed there, the most common name is used (e.g. the designation in the 'Catalogue of Suspected Variables'

(Kholopov, 1982)). Provisional designations and numbers in general star catalogues are given in parentheses. The provisional names, whose continued use is discouraged, are often useful for identifying papers referring to the nova before the definite designation was assigned.

The type of variability is that given in the GCVS, or an extended classification according to the following scheme:

- NA – fast nova (t_3-time < 100^d), spectroscopically confirmed;
- NB – slow nova (t_3-time ≥ 100^d), spectroscopically confirmed;
- NC – extremely slow nova (typical time scales: decades);
- N – nova; light curve too poorly known to establish the speed class.

The following additions are used:

- : – nova, not confirmed by spectroscopic observations;
- ? – existence of object based on one photographic observation or a few independent visual observations;
- ?? – existence based on no more than two visual observations.

Other types of variability:

- NR – recurrent nova;
- NL – novalike variable (inhomogeneous group);
- UG – U Gem variable (dwarf nova);
- UGSS – U Gem variable (SS Cyg subtype);
- UGSU – U Gem variable (SU UMa subtype);
- UGZ – U Gem variable (Z Cam subtype);
- UGWZ – U Gem variable (WZ Sge subtype);
- ZAND – symbiotic variable (Z And type);
- XND – X-ray nova-like (transient) system consisting of a compact component and a dwarf or subgiant star with spectral type G to M;
- M – Mira star;
- UV – UV Ceti type star;
- SDOR – S Dor type star.

Discovery

A short report on the circumstances of discovery is given, followed by a bibliographical note on the discovery. In connection with magnitudes, '[' indicates 'fainter than' and ']' indicates 'brighter than'.

Positions

All listed positions refer to the equinox 1950.0.

Positions are given in the following (unnumbered) sequence:

1. Equatorial coordinates (h, m, s, °, ′, ″) of the pre- or exnova as measured on a glass copy of the Palomar Observatory Sky Survey (POSS) or the ESO/SRC Sky Atlas (SRC); in some cases SERC Equatorial Atlas plates were used. The objects were

Nova Catalogue	IAU Standard Abbreviation (place of issue)
DDO Comm	Commun. David Dunlap Obs. (Toronto)
DDO Publ	Publ. David Dunlap Obs. (Toronto)
Erg AN	Ergänzungshefte der Astr. Nachr. (Berlin)
HA	Ann. Harvard Coll. Obs. (Cambridge/USA)
HAC	Harvard Coll. Obs. Announc. Cards (Cambridge/USA)
Harv Repr	Harvard Reprints (Cambridge/USA)
HB	Bull. Harvard Coll. Obs. (Cambridge/USA)
HC	Circ. Harvard Coll. Obs. (Cambridge/USA)
Heidelberg Ver	Veröff. Sternw. Heidelberg (Karlsruhe; Heidelberg)
IAU Circ	Circ. Centr. Bur. astr. Telegr. (Cambridge/USA)
IBVS	Comm. 27 IAU Inf. Bull. var. Stars (Budapest)
Izv KRAO	Izv. Krym. astrofiz. Obs. (Moskva)
JBAA	J. Br. astr. Ass. (London)
JO	J. Obs. (Marseille)
JRAS Can	J. R. astr. Soc. Can. (Toronto)
København Publ	Publ. mind. Medd. Københavns Obs. (København)
KOB	Kodaikanal Obs. Bull. (Madras)
KVB	Kl. Veröff. Remeis-Sternw. (Bamberg)
KVBB	Kl. Veröff. Sternw. Berlin-Babelsberg (Berlin)
Leiden Ann	Ann. Sterrew. Leiden (Leiden)
Lick Bull	Lick Obs. Bull. (Berkeley; Santa Cruz; Los Angeles)
Lick Publ	Publ. Lick Obs. (Berkeley)
Louvain Publ	Publ. Lab. Astr. Géod. Univ. Louvain (Louvain)
Lund Ann	Ann. Obs. Lund (Stockholm)
Lwów Contr	Contr. astr. Inst. Lwów Univ. (Lwów)
Lyon Publ	Publ. Obs. Lyon (Lyon)
Mem RAS	Mem. R. astr. Soc. (London)
Mem SA It	Mem. Soc. astr. Ital. (Roma)
Michigan Publ	Publ. Obs. Univ. Michigan (Ann Arbor)
MN	Mon. Not. R. astr. Soc. (London)
MNASSA	Mon. Not. R. astr. Soc. South. Afr. (Cape Town)
MB PrAW	Monatsber. K. Preuß. Akad. Wiss. Berlin (Berlin)
MVS	Mitt. veränderl. Sterne (Sonneberg)
Nature	Nature (London)
Nbl AZ	Nachrichtenblatt astr. Zentralstelle (Heidelberg; Karlsruhe)
NZAS Publ	Publ. Var. Star Sect. R. astr. Soc. New Zealand (Greerton)
Obs	Observatory (London)
PA	Pop. Astr. (Northfield)
PAAS	Publ. am. astr. Soc.
PASJ	Publ. astr. Soc. Japan (Tokyo)
PASP	Publ. astr. Soc. Pacific (San Francisco)
Phil Trans	Phil. Trans. R. Soc. London (London)
Pisma AZh	Pisma Astr. Zhurnal (Moskva)
Proc R Soc	Proc R. Soc. (London)
PZv	Perem. Zvezdy (Nishni-Novgorod; Moskva-Leningrad)
QJRAS	Q. J. R. astr. Soc. (London)
Ric Astr	Ric. astr. Specola astr. Vatic. (Città del Vaticano)
Riv Publ	Riverview Coll. Obs. Publ. (Riverview, NSW)
Riv Repr	Riverview Coll. Obs. Reprints (Riverview, NSW)
SAAO Circ	S. Afr. Astr. Obs. Circ. (Observatory, Cape)
SAO SpR	Smithsonian astrophys. Obs. Special Report (Cambridge/USA)
Sov Astr	Sov. Astr. (New York)

Nova Catalogue	IAU Standard Abbreviation (place of issue)
Sov Astr Lett	Sov. Astr. Lett. (New York)
Space Sci Rev	Space Sci. Rev. (Dordrecht)
Sterne	Sterne (Leipzig)
Stockh Ann	Stockholms Obs. Ann. (Stockholm)
SuW	Sterne und Weltraum (Mannheim)
Tadjik Bull	Bjull. Stalinabad (Dushanbe) astr. Obs. (Dushanbe)
Tadjik Tsirk	Circ. Tadjik astr. Obs. (Dushanbe)
Tokyo Bull	Tokyo astr. Bull (Tokyo)
Ton Bol	Bol. Inst. Tonantzintla (Puebla)
TBB	Bol. Obs. Tonantzintla y Tacubaya (Mexico)
UOC	Union Obs. Circ. (Johannesburg)
Victoria Publ	Publ. Dom. astrophys. Obs. Victoria (Victoria)
Vistas	Vistas in Astronomy (Oxford)
VJS	Vierteljahresschr. astr. Ges. (Leipzig)
VSS	Veröff. Sternw. Sonneberg (Berlin)
Wien Mitt	Mitteilungen der Universitäts-Sternwarte Wien (Wien)
ZsAp	Z. Astrophys. (Berlin; Göttingen: Heidelberg)

7. Prospects

It is my sincere hope that the catalogue and atlas may prove useful in observations and studies of the novae listed here, that the number of errors and misidentifications is not too large, and that all shortcomings noted by the users will be reported to the author.

It is planned to publish from time to time supplements and revisions. Their extent will depend on all future observers of old and new novae.

Acknowledgements

Many colleagues helped me to carry out this project. First of all, my thanks go to Waltraut Seitter, who helped me in all stages; she copied all the Harvard plates, participated in all observations at the La Silla and Calar Alto Observatories, and offered to work for me at the Astronomical Institute at Münster. Without her, this project would never have come to a successful end. It is my pleasure to dedicate this catalogue and atlas to her.

I also thank Martha Hazen, Curator of Astronomical Photographs at the Harvard College Observatory, Cambridge (USA), for her support in this project. From the Harvard plates, the majority of the new identifications given here were obtained. I also thank C. Y. Shao, Center for Astrophysics, for the loan of some plates.

Emilia Belserene, Director of the Maria Mitchell Observatory, Nantucket (USA), kindly offered the use of the MMO plate collection for outburst studies. The task was carried out by A. Sarajedini with great skill.

Further help came from G. Klare and J. Krautter, Landessternwarte Heidelberg, R. Knigge, Remeis-Sternwarte Bamberg, the late A. Przybylski, Siding Spring Observatory, W. Pfau, Universitätssternwarte Jena, H. Debehogne, Observatoire Royal de Bruxelles, H. Kosai, Tokyo Astronomical Observatory, R. Argyle, Royal Greenwich Observatory/La Palma Observatory, S. Dieters, University of Tasmania, Hobart. Bibliographical information was supplied by M. F. Bode, Lancashire Polytechnic, Preston, and by C. Jaschek, Centre des Données Stellaires, Strasbourg. I also thank for the support given by R. Budell, R. Duemmler and H.-J. Tucholke, Münster.

I thank Professor C. de Jager, for his agreement to accept this catalogue for Space Science Reviews. The fine cooperation of the publishers, especially Messrs. G. Kiers, G. Oomen, and F. van Schaik and their colleagues, is gratefully acknowledged.

I thank the California Institute of Technology and the Royal Observatory Edinburgh for permission to reproduce sky atlas plates.

The stay at the Harvard College Observatory and the observing runs at the Calar Alto Observatory were supported by travel grants of the Deutsche Forschungsgemeinschaft.

Finally, I should like to express my deep respect for the innumerable known and unknown astronomers and night assistants who exposed millions of plates since the early days of photography, a minute part of which was used for the production of this catalogue and atlas. Extending the motto quoted from Pliny: these observers are truly Hipparchos' disciples to whom we owe so much.

General References

Allen, D. A.: 1984, *Proc. Astron. Soc. Australia* **5**, 369.

Bertaud, C.: 1945, *Ann. Obs. Paris* **9**, 1.

BD – F. W. A. Argelander: 1863, *Bonner Durchmusterung des nördlichen Sternhimmels.*

Bruch, A., Fischer, F. J., and Wilmsen, U.: 1987, *Astron. Astrophys.* (in preparation).

Cecchini, G., and Gratton, L.: 1941, *Le Stelle Nuove*, Pubbl. R. Oss. Milano-Merate = Memorie della Reale Accademia d'Italia, Classe di Scienze fisiche, matematische e naturali, ser. 6, fasc. 1.

Duerbeck, H. W.: 1981, *Publ. Astron. Soc. Pacific* **93**, 165.

Duerbeck, H. W. and Seitter, W. C.: 1979, *ESO Messenger*, No. 17, 1.

Efremov, Yu. N.: 1961, *Perem. Zvezdy* **13**, 317.

GCVS4 – see Kholopov (1985).

Humason, M.: 1938, *Astrophys. J.* **88**, 228.

Kenyon, S. J.: 1986, *The Symbiotic Stars*, Cambridge, Cambridge Univ. Press.

Khatisov, A. Sh.: 1971, *Byull. Abastumani Astrofiz. Obs.* **40**, 13.

Kholopov, P. N.: 1982, *New Catalogue of Suspected Variable Stars*, Moskva, Nauka.

Kholopov, P. N.: 1985, *General Catalogue of Variable Stars*, Moskva, Nauka.

Li, J., Xiang, X., and Liu, X.: 1984a, *Publ. Beijing Astron. Obs.*, No. 4, 1.

Li, J., Xiang, X., and Liu, X.: 1984a, *Publ. Beijing Astron. Obs.*, No. 6, 87.

Liu, X. and Li, J.: 1984, *Publ. Yunnan Obs.*, No. 1, 1.

Müller, G. and Hartwig, E.: 1918, *Geschichte und Literatur des Lichtwechsels*, Vol. 1–3, Leipzig, Poeschel und Trepte; continued by R. Prager and H. Schneller, *Geschichte und Literatur des Lichtwechsels veränderlicher Sterne*, 2. Ausgabe, Berlin, F. Dümmler and Akademie-Verlag, 1934ff.

NSV – see Kholopov (1982).

Payne-Gaposchkin, C.: 1957, *The Galactic Novae*, Amsterdam, North-Holland Publ. Co.

Payne-Gaposchkin, C.: 1977, in M. Friedjung (ed.), *Novae and Related Stars*, Dordrecht, D. Reidel Publ. Co., p. 1.

Perek, L. and Kohoutek, L.: 1967, *Atlas of Galactic Planetary Nebulae*, Praha, Academia Publ. House.

Ritter, H.: 1984, *Astron. Astrophys. Suppl.* **57**, 385.

Seitter, W. C. and Duerbeck, H. W.: 1986, in M. F. Bode (ed.), *RS Ophiuchi and the Recurrent Nova Phenomenon*, Utrecht, VNU Press, p. 71.
Tsesevich, V. P. and Kazanasmas, M. S.: 1971, *Atlas of Finding Charts of Variable Stars*, Moskva, Nauka.
Vogt, N. and Bateson, F.: 1982, *Astron. Astrophys. Suppl.* **48**, 383.
Wenzel, W.: 1981, *Bull. Inf. CDS*, No. 20, 105.
Wenzel, W. and Huth, H.: 1981, *Bibliography of Variable Stars* (microfiche edition, CDS Strasbourg).
Williams, G.: 1983, *Astrophys. J. Suppl.* **53**, 523.
Wyckoff, S. and Wehinger, P. A.: 1978, *Publ. Astron. Soc. Pacific* **90**, 557.

THE CATALOGUE

All listed positions refer to the equinox 1950.0. For detailed explanations on arrangements, abbreviations and symbols, consult Section 5, page 5.

LS And — NA:

(NSV 00191)

Discovered by E. Herbst on a Palomar Schmidt plate taken 1971 August 26 (S. van den Bergh, E. Herbst, C. Pritchet, *AJ* **78** (1973) 275).

Position:	00 29 28.42	+41 41 39.2	(POSS)
	00 29 28.55	+41 41 37.9	(A. S. Sharov, D. K. Karimova, *ATs* 998 (1978) 1)
	119.104	−20.786	(G.C.)

Range: **11.7**p – 20.5p LCT: A t_3: 8 d

Finding chart: S. van den Bergh, E. Herbst, C. Pritchet, *AJ* **78** (1978) 275.
Light curve: A. S. Sharov, D. K. Karimova, *ATs* 998 (1978) 1.
Identification: from A. S. Sharov and D. K. Karimova's position.
Classification: intergalactic very fast nova? A faint eruptive star at high galactic latitude with the light curve of a very fast nova. No spectroscopic information is available.

N And 1986 — N

Discovered by M. Suzuki, Japan, on 1986 December 5, as a star of $8^{m}_{\cdot}0$. It was already 8^{m} on December 4, invisible on December 2 (*IAU Circ* 4281).

Position:	23 09 47.47	+47 12 00.6	(POSS)
	23 09 47.66	+47 12 00.8	(2 outburst observations)
	106.052	−12.117	(G.C.)

Range: 6.3v – 17.8p LCT: A t_3: 22 d

Spectroscopy: Y. Norimoto, *IAU Circ* 4281 (1986) descr; C. C. Huang, *IAU Circ* 4286 (1986) – descr.
Identification: from published precise positions. Finding chart in Appendix.
Classification: nova.

VY Aqr UGWZ

(N Aql 1907, 1907.6 Aqr, Ross 88)

Discovered by F. E. Ross, Yerkes Observatory, on a plate taken 1907 August 12 (*AJ* **36** (1925) 123).

Position:	21 09 28.26	– 09 01 56.4	(POSS)
	21 09 28.36	– 09 01 51.1	(R. W. Argyle, *IAU Circ* 2896 (1983))
	41.593	– 35.214	(G.C.)

Range: 8.4p – 17.2p LCT: DN t_3: variable

Finding chart:	S. Wyckoff, P. A. Wehinger (1978).
Light curve:	I. E. Woods, *HB* 836 (1925); M. Verdenet, E. Schweitzer, *BAFOEV* **27** (1984) 6.
Spectroscopy:	E. M. Hendry, *IBVS* 2381 (1983) – minimum spectrum, descr, rv.
Duplicity:	spectroscopic binary with $P = 0.22^d$ (H. Ritter 1984).
Identification:	from Wyckoff and Wehinger's finding chart.
Classification:	dwarf nova with long cycle length, previously classified as recurrent nova. Major outbursts in 1907, 1929, 1934, 1941?, 1942, 1958, 1973, 1983, 1986, with maxima of different height. E. M. Hendry (*IBVS* 2381 (1983)) suggests recurrence period of ~11.0 years.

CI Aql UG or N?

(23.1925 Aql)

Discovered by K. Reinmuth on Heidelberg plates of 1917 June 25 (*AN* **225** (1925) 385).

Position:	18 49 28.03	– 01 32 19.1	(POSS)
	18 49 28.20	– 01 32 24.6	(K. Reinmuth, *AN* **225** (1925) 385)
	31.688	– 0.812	(G.C.)

Range: 11.0p – 15.6p LCT: ? t_3: ?

Light curve:	K. Reinmuth, *AN* **225** (1925) 385; P. P. Parenago, *PZv* **3** (1931) 99.
Identification:	from Heidelberg plates B 3961/62, taken 1917 June 25.
Classification:	unclear. Small outburst amplitude suggests either dwarf nova with long cycle length or nova outburst whose maximum was missed. No spectroscopic information is available.

DO Aql **NB**

(N Aql 1925, 9.1925 Aql)

Discovered by M. Wolf on Heidelberg plates taken 1925 September 14 (*AN* **225** (1925) 335). The nova is first seen on a Harvard plate, taken 1925 September 8, as a star of $9^{m}.8$p.

Position:			
	19 28 45.10	− 06 32 02.4	(POSS)
	19 28 44.95	− 06 32 02.3	(M. Wolf, *AN* **225** (1925) 335)
	31.705	− 11.805	(G.C.)

Range: **8.7**v − 16.5p LCT: D t_3: 900^d

Finding chart: Yu. N. Efremov (1961).
Light curve: M. Beyer, *AN* **235** (1929) 427; H. Vogt, *AN* **232** (1928) 269; C. Payne-Gaposchkin (1957) 163; G. Cecchini, L. Gratton (1941) 133.
Spectroscopy: A. J. Cannon, *HB* 826 (1925) – descr; N. Tikhoff, *AN* **228** (1926) 296 – descr; P. W. Merrill, *PASP* **38** (1926) 387 – phot; B. Vorontsov-Velyaminov, *ApJ* **92** (1940) 283 – discussion of Merrill's spectra.
Identification: from Heidelberg plates B 5038/5039, taken 1925 September 19.
Classification: extremely slow nova with protracted flat-topped maximum.

EL Aql **NA**

(N Aql 1927, 60.1927 Aql)

Discovered by M. Wolf on Heidelberg plates taken 1927 July 30 and 31 (*AN* **230** (1927) 421). On Harvard plates, the nova is [$11^{m}.1$ on 1927 June 8, $6^{m}.4$ on 1927 June 15.

Position:			
	18 53 24.27	− 03 23 15.95	(POSS)
	18 53 24.60	− 03 23 15.05	(3 outburst observations)
	30.497	− 2.535	(G.C.)

Range: **6.4**p − 20p LCT: Ba t_3: 25^d

Finding chart: M. Wolf, *AN* **230** (1927) 422.
Light curve: A. J. Cannon, *HB* 851 (1927) 9; M. Harwood, *HB* 851 (1927) 10; J. Voûte, *BAN* **4** (1927) 106; M. Beyer, *AN* **235** (1929) 427; C. Payne-Gaposchkin (1957) 9,165; G. Cecchini, L. Gratton (1941) 138, 139.

Spectroscopy: M. Humason, *PASP* **39** (1927) 369 – phot; G. Shain, W. Nikonoff, *AN* **233** (1928) 222 – trac; A. B. Wyse, *Lick Publ* **14** (1940) 217 – ident, rv.

Identification: the discovery plate, Heidelberg Tessar 795/796, has too small a scale. Harvard plates MC 22619, taken 1927 August 3/4, and MC 22672, taken 1927 September 8/9, were used.

Classification: well-observed fast nova.

EY Aql NA:

(17.1929 Aql, SVS 202)

Discovered by V. Albitzky on plates of the Simeis Observatory. The object is seen on 5 plates between 1926 September 8 and 30 (*AN* **235** (1929) 317).

Position:	19 32 27.23	+ 14 55 13.7	(POSS)
	19 32 27	+ 14 55 06	(V. Albitzky, *AN* **235** (1929) 317)
	51.222	− 2.476	(G.C.)

Range: 10.5p – 20r/[21p LCT: Cb? t_3: 40 d

Finding chart: V. Albitzky, *AN* **235** (1929) 317; Yu. N. Efremov (1961) – not identified.

Light curve: H. W. Duerbeck, *IBVS* 2490 (1984).

Identification: Harvard plates MF 10709, taken 1926 September 2/3, and MF 10885, taken 1926 October 1/2, were used. The nova is faintly visible on the POSS red plate, it is below the limit of the POSS blue plate.

Classification: faint, poorly observed moderately fast nova without spectroscopic confirmation.

V352 Aql ZAND

(279.1931 Aql, P1753, PN 37-3.3, K3-25)

Discovered on Sonneberg plates of 1928 (C. Hoffmeister, E. Ahnert, W. Götz, H. Huth, P. Ahnert, *VSS* **2** (1954) 52).

Position:	19 11 02.33	+ 02 13 02.6	(SRC)
	19 11 05	+ 02 13	(C. Hoffmeister, E. Ahnert, W. Götz, H. Huth, P. Ahnert, *VSS* **2** (1954) 52)
	37.513	− 3.863	(G.C.)

Range: 13.3p – 18B LCT: D? t_3: $>2000^d$

Finding chart: *MVS* 250 (1957); L. Perek, L. Kohoutek (1967).
Light curve: C. Hoffmeister, E. Ahnert, W. Götz, H. Huth, P. Ahnert, *VSS* **2** (1954) 52.
Identification: from Perek and Kohoutek's finding chart, confirmed by spectroscopic observation.
Classification: symbiotic star (H. W. Duerbeck, W. C. Seitter, *MN* (in press)).

V356 Aql — NB

(N Aql 1936 No. 1, 618.1936 Aql)

Discovered by N. Tamm, Kvistaberg Observatory, on 1936 September 18 (*AN* **260** (1936) 375).

Position:	19 14 41.69	+01 37 56.0	(SRC)
	19 14 41.79	+01 37 56.2	(15 outburst observations)
	37.419	− 4.944	(G.C.)

Range: **7.7**p – 17.7p LCT: D t_3: 115^d

Finding chart: Yu. N. Efremov (1961).
Light curve: P. P. Parenago, *PZv* **7** (1949) 109; C. Hoffmeister, R. Morgenroth, *AN* **260** (1936) 368; C. Payne-Gaposchkin (1957) 15,168; G. Cecchini, L. Gratton (1941) 161.
Spectroscopy: J. Stein, A. Zirwes, *AN* **260** (1936) 388 – premaximum spectrum; F. Hinderer, *AN* **260** (1936) 387 – descr, rv; D. B. McLaughlin, *ApJ* **122** (1955) 417 – descr.
Duplicity: d = 0″.18, PA 140° (1936.75) (H. E. Wood, *MN* **97** (1937) 320).
Identification: from Efremov's finding chart.
Classification: quite slow nova with prolonged maximum, showing fluctuations of 1–2^m amplitude.

V368 Aql — NA

(N Aql 1936 No. 2, N Aql No. 6, 668.1936 Aql)

Discovered by N. Tamm at Kvistaberg Observatory on a photographic plate taken 1936 October 7 (*AN* **261** (1936) 15). Maximum light occurred 1936 September 25, followed by a rapid decline.

Position:	19 24 08.97	+07 30 09.0	(6 recent observations)
	19 24 09.06	+07 30 08.9	(7 outburst observations)
	43.728	− 4.265	(G.C.)

Range: 6.55p (6.1) – 17.8:p LCT: Ao t_3: 42^d

Finding chart: Yu. N. Efremov (1961).
Light curve: S. Gaposchkin, *HB* 917 (1943) 16; P. Ahnert, *Sterne* **23** (1943) 16; C. Hoffmeister, *BZ* **25** (1943) 104; C. Payne-Gaposchkin (1957) 11.
Spectroscopy: R. F. Sanford, *PASP* **55** (1943) 284 – descr; G. Williams (1983) – minimum spectrum, trac.
Identification: Heidelberg plates B 6920/6021, taken 1943 October 29, Harvard plate MC 33177, taken 1943 November 25/26, confirm Efremov's identification.
Classification: typical moderately fast nova, fairly well documented.

V500 Aql NA

(N Aql 1943, 215.1943 Aql, S 3542)

Discovered by C. Hoffmeister on Sonneberg plates. The nova reached maximum light between 1943 April 13 ([13^{m}5p) and 1943 May 2 (6^{m}55p) (*IAU Circ* 961, *BZ* **25**, 104).

Position: 19 50 02.89 +08 20 58.9 (POSS)
19 50 03.07 +08 20 59.6 (H. Krumpholz, *BZ* **25** (1943) 124)
47.608 – 9.462 (G.C.)

Range: 6.55p (6.1) – 17.8p LCT: Ao t_3: 42^d

Finding chart: Yu. N. Efremov (1961).
Light curve: S. Gaposchkin, *HB* 917 (1943) 16; P. Ahnert, *Sterne* **23** (1943) 16; C. Hoffmeister, P. Ahnert, *BZ* **25** (1943) 104; C. Payne-Gaposchkin (1957) 11.
Spectroscopy: R. F. Sanford, *PASP* **55** (1943) 284 – descr.
Identification: Heidelberg plates B 6920/6921, taken 1943 October 29, and Harvard plates MC 33177, taken 1943 November 25/26, MC 33193, taken 1943 November 29/30, confirm Efremov's identification.
Classification: moderately fast nova.

V528 Aql NA

(N Aql 1945)

Discovered by C. Bertaud, Meudon, 1945 August 26 (*IAU Circ* 1014), and independently by N. Tamm, Kvistaberg Observatory, 1945 August 28.

Position: 19 16 45.93 +00 32 19.2 (SRC)
19 16 46.00 +00 32 18.5 (6 outburst observations)
36.687 – 5.910 (G.C.)

Range: **7.0**p – 18.1p LCT: Ao or Ba t_3: 37^d

Finding chart: J. Stein, W. J. Miller, *Ric Astr* **2** (1948) 49; S. Taffara, *Mem SA It* **35** (1964) 125.
Light curve: O. D. Dokuchaeva, *PZv* **7** (1949) 95; A. Hagopian, C. B. Sawyer, *HB* 918 (1946) 5; C. Schalén, A. Wallenquist, *AMAF* **33A** (1946) 1; J. Stein, W. J. Miller, *Ric Astr* **2** (1948) 49; C. Bertaud, *JO* **34** (1951) 38; C. Payne-Gaposchkin (1957) 11, 172.
Spectroscopy: F. J. Neugebauer, G. H. Herbig, *PASP* **57** (1945) 264 – phot; R. F. Sanford, *PASP* **57** (1945) 263,321 – descr, phot; J. Stein, W. J. Miller, *Ric Astr* **2** (1948) 49 – phot, trac, ident; D. B. McLaughlin, *AJ* **58** (1953) 220 – rv, descr; D. B. McLaughlin, *ApJ* **131** (1960) 739 – descr; D. B. McLaughlin, *AAp* **27** (1964) 450 – rv.
Duplicity: companion 16.5 at d = 11.″5, PA 59°.
Identification: from Harvard plate IR 7904, taken 1945 August 30/31.
Classification: well-observed fast nova.

V603 Aql NA

(N Aql 1918, N Aql No. 3, 7.1918 Aql, HD 174107)

Discovered by G. N. Bower and others 1918 June 8/9, when the nova was 1^m. On the evening of June 7, the nova is 6^m on a Harvard plate (*AN* **207** (1918) 17).

Position:	18 46 21.45	+ 00 31 36.1	(3 new observations)
	18 46 21.45	+ 00 31 37.05	(12 outburst observations)
	13.164	+ 0.829	(G.C.)

Range: **– 1.1**v – 12.0v LCT: Ao t_3: 8^d

Finding chart: M. Humason (1938); A. Sh. Khatisov (1971); G. Williams (1983).
Light curve: L. Campbell, *HA* **81** (1919) 113; H. Shapley, *HA* **81** (1922) 239; C. Bertaud (1945) 62; C. Payne-Gaposchkin (1957) 9, 85; G. Cecchini, L. Gratton (1941) 106.
Spectroscopy: A. J. Cannon, *HA* **81** (1920) 179 – phot, ident, descr, line intensities; A. B. Wyse, *Lick Publ* **14** (1940) 93 – trac, phot, ident, descr; C. Payne-Gaposchkin, S. Gaposchkin, *HC* 145 (1942) – spectrophotometry; C. Bertaud (1945) – tables; C. Payne-Gaposchkin (1957) 85; G. Cecchini, L. Gratton (1941) 107, 108; E. S. Pearson, *ASPO Camb* **4** (1936) 85 – phot, ident, rv; J. Lunt, *MN* **79** (1918) 418, *MN* **80** (1919) 519, *MN* **80** (1920) 696 – phot, ident, rv; J. Evershed, *MN* **79** (1918) 468 – phot, ident, rv; F. E. Baxandall, *MN* **81** (1920) 66 – phot, ident, rv; G. Williams (1983) – minimum

	spectrum, trac; R. P. Kraft, *ApJ* **139** (1964) 457 – minimum spectrum, phot.
Nebular shell:	E. E. Barnard, *ApJ* **49** (1919) 199; A. B. Wyse, *Lick Publ* **14** (1940) 93; H. F. Weaver, *Highlights of Astronomy* **3** (1974) 509; H. W. Duerbeck, *ApSS* **131** (1987) 461.
Duplicity:	spectroscopic binary, possibly eclipsing, P = 0.13854d (H. Ritter 1984).
Identification:	from published finding charts.
Classification:	well-observed very fast nova.

V604 Aql **NA**

(N Aql 1905, N Aql No. 2, 104.1905 Aql, HV 1175, HD 176779)

Discovered by W. Fleming on Harvard plates taken on 1905 August 31. The nova is [$9^{m}.7$ on 1905 August 10, $8^{m}.2$ on August 17, rapidly fading (*HB* 197 (1905), *AN* **169** (1905) 207).

Position:	18 59 27.47	– 04 31 07.5	(POSS)
	18 59 27.82	– 04 31 07.1	(6 outburst observations)
	30.182	– 4.395	(G.C.)

Range: 8.2p (7.6) – 21p LCT: A t_3: 25^d

Finding chart:	E. C. Pickering, *HC* 106 (1905) 1; A. D. Walker, M. Olmsted, *PASP* **70** (1958) 495; S. Wyckoff, P. A. Wehinger (1978) give wrong identification and coordinates.
Light curve:	A. D. Walker, *HA* **84** (1923) 189; C. Payne-Gaposchkin (1957) 9; G. Cecchini, L. Gratton (1941) 78.
Spectroscopy:	J. H. Moore, *ApJ* **23** (1906) 261 – descr; A. J. Cannon, *HA* **76** (1916) 19 – descr.
Identification:	from Heidelberg plates B1301/1302, taken 1905 September 17.
Classification:	fast nova; in obscured region of the Galaxy.

V605 Aql **NB (pec)**

Discovered by M. Wolf on Heidelberg plates taken 1919 July 4 (*AN* **211** (1919) 119).

Position:	19 15 48.67	+ 01 41 28.8	(SRC)
	19 15 48.7	+ 01 41 33.8	(M. Mündler, *AN* **211** (1919) 119)
	37.601	– 5.163	(G.C.)

Range: **10.4**v – 22.5v LCT: D t_3: 1500^d

Finding chart: M. Wolf, *AN* **211** (1919) 119; G. H. Herbig, *PASP* **70** (1958) 605; S. van den Bergh, *PASP* **83** (1971) 819.

Light curve: W. C. Seitter, *AG Mitt* **63** (1985) 181; W. C. Seitter, *Sterne* (1987) – in press.

Spectroscopy: K. Lundmark, *PASP* **33** (1921) 314, *PASP* **34** (1922) 210 – descr; W. C. Seitter, *ApJ* (in press).

Nebular shell: W. C. Seitter, in ESO Workshop on Production and Distribution of C,N,O Elements (1985), ed. J. Danziger, p. 253.

Identification: from Harvard plate MF 4617, taken 1919 August 26, and Heidelberg plates B 4896/4897, taken 1924 July 7.

Classification: peculiar nova with C spectrum at maximum; hydrogen-poor nebular remnant and WR/O VI type central star at minimum, central object of the planetary nebula A58. Final He-flash of nucleus of planetary nebula? (W. C. Seitter, *Sterne* (1987)).

V606 Aql NA

(N Aql 1899, N Aql No. 1, 11.1900 Aql, HV 150, HD 181419, BD $-0°3708^{a}$)

Discovered by W. Fleming on Harvard plates. The nova is first seen on a plate taken 1899 April 21. The last plate before outburst was taken 1898 November 1 ([13^{m}]) (*AN* **153** (1900) 59).

Position:			
(1)	19 17 50.07	− 0 13 40.7	(POSS)
(2)	19 17 50.365	− 0 13 40.8	(POSS)
(3)	19 17 49.85	− 0 13 48.3	(POSS)
	19 17 50.30	− 0 13 41.1	(A. Sh. Khatisov (1971))
	36.128	− 6.502	(G.C.)

Range: 6.7p (5.5) – 17.3p LCT: ? t_3: 65^{d}

Finding chart: A. D. Walker, M. Olmsted, *PASP* **70** (1958) 495; Yu. N. Efremov (1961); A. Sh. Khatisov (1971); S. Wyckoff, P. A. Wehinger (1978).

Light curve: H. Leavitt, *HA* **84** (1920) 121; G. Cecchini, L. Gratton (1941) 59.

Spectroscopy: A. J. Cannon, *HA* **76** (1916) 19 – descr.

Identification: from Harvard plates A 6012, taken 1902 July 9, and B 22707, taken 1899 May 2. The exnova is a blend of three stars which are listed above.

Classification: nova whose early stages are not covered by observations. The light curve shows a steep decline, followed by a plateau of 100^{d} duration.

V607 Aql **M:**

(N Aql 1904, 31.1905 Aql, Ross 219, Zi 1733)

Discovered by M. and G. Wolf on Heidelberg plates taken 1904 July 6 (*AN* **167** (1905) 337), independently by F. E. Ross (*AJ* **37** (1927) 155) on a plate taken 1904 June 21.

Position:	19 32 52.42	+ 0 36 01.3	(SRC)
	19 32 52.3	+ 0 36 01.6	(M. and G. Wolf, *AN* **167** (1905) 337))
	38.635	− 9.445	(G.C.)

Range: 11.5p – [18.5p LCT: M? t_3: –

Finding chart:	M. Wolf, G. Wolf, *AN* **167** (1905) 337.
Identification:	from Wolf and Wolf's chart and position.
Classification:	H. U. Sandig (*AN* **279** (1950) 89) first classified this object as nova. G. Richter (*MVS* 554 (1961)) found Mira type variability ($14^m.1$p – [$18^m.5$p, P = 474^d). The 1904 maximum might have been a bright Mira maximum; symbiotic behaviour cannot be excluded. More observations are needed to improve the classification.

V841 Aql **N**

(N Aql 1951)

Discovered by F. Zwicky, Mt. Palomar Observatory, on an objective prism plate taken 1951 July 10 (*IAU Circ* 1319, *HAC* 1130). The nova appears on Harvard plates as early as 1951 April 16 ($12^m.1$).

Position:	19 05 17.92	+ 10 24 57.1	(POSS)
	19 05 17	+ 10 25 48	(F. Zwicky, *IAU Circ* 1319 (1951))
	44.111	+ 1.212	(G.C.)

Range: 11.5p – 17.5p LCT: ? t_3: ?

Finding chart:	Yu. N. Efremov (1961).
Light curve:	D. Hoffleit, *HB* 921 (1952) 4; C. Hoffmeister, *NB1 AZ* **5** (1951) 32.
Spectroscopy:	F. Zwicky, R. Minkowski, *HAC* 1130 (1951).
Identification:	from Efremov's finding chart. On POSS chart 506, taken 1952 May 24, the nova is in decline.
Classification:	nova whose early outburst stages are not covered by observations. The light curve shows slow decline with superimposed fluctuations.

V890 Aql –

(N Aql 1946)

Observed 1946 August 30 by R. Rigollet (*Astronomie* **61** (1947) 189); confusion with minor planet 258 Tyche (N. N. Samus, *ATs* 1215 (1982) 8).

Classification: nova does not exist; no finding chart is given.

V1229 Aql NA

(N Aql 1970)

Discovered by M. Honda, Japan, 1970 April 14/15 (*IAU Circ* 2233).

Position:			
	19 22 15.41	+ 04 08 50.4	(POSS)
	19 22 15.40	+ 04 08 49.6	(5 outburst observations)
	40.537	− 5.438	(G.C.)

Range: **6.7**v – 19.4 p LCT: A? B? t_3: 37^d

Finding chart: A. Sh. Khatisov (1971); H. Kosai, *Tokyo Bull 2nd Ser* **214** (1971) 2515.
Light curve: F. Ciatti, L. Rosino, *AsAp Suppl* **16** (1974) 305; I. D. Howarth, *JBAA* **88** (1978) 180.
Spectroscopy: F. Ciatti, L. Rosino, *AsAp Suppl* **16** (1974) 305 – phot. ident; T. C. Grenfell, *PASP* **83** (1971) 66.
Identification: from Harvard plate IR 12488, taken 1970 April 15/16.
Classification: fast nova.

V1301 Aql NA

(N Aql 1975)

Discovered by P. Wild, Zimmerwald Observatory, on a plate taken 1975 June 4 (*IAU Circ* 2788).

Position:			
	19 15 26.75	+ 04 41 49.3	(POSS)
	19 15 26.9	+ 04 41 44.3	(P. Wild, *IAU Circ* 2788 (1975))
	40.225	− 3.681	(G.C.)

Range: **10.3**v – 21p LCT: A? B? t_3: 35^d

Finding chart: I. D. Howarth, *MVS* **7** (1976) 110.
Light curve: J. A. Mattei, *JRAS Can* **69** (1975) 319; I. D. Howarth, *MVS* **7** (1976) 110; F. J. Vrba, G. D. Schmidt, E. W. Burke, Jr., *ApJ* **211** (1977) 480.
IR observations: F. J. Vrba, G. D. Schmidt, E. W. Burke, Jr., *ApJ* **211** (1977) 480.
Spectroscopy: P. Pesch, *IAU Circ* 2835 (1975).
Identification: from Harvard plate SH 4973, taken 1975 July 4/5.
Classification: fast nova.

V1333 Aql — XND

(Aql X-1, 4U1908 + 00)

Optical counterpart of the flaring X-ray source Aql X-1, flares with time scales of hundreds of days.

Position:	19 08 42.83	+ 0 30 05.5	(SRC)
	19 08 42.9	+ 0 30 05	(J. Thorstensen, P. Charles, S. Bowyer, *ApJ* **220** (1978) L131)
	35.719	− 4.143	(G.C.)

Range: 14.8v – 19.4v LCT: ? t_3: ?

Finding chart: J. Thorstensen, P. Charles, S. Bowyer, *ApJ* **220** (1978) L131.
Light curve: J. Thorstensen *et al.*; P. A. Charles, J. R. Thorstensen, S. Bowyer, G. W. Clark, F. K. Li, J. van Paradijs, R. Remillard, S. S. Holt, L. J. Kaluzienski, V. T. Junkkarinen, R. C. Puetter, H. E. Smith, G. S. Pollard, P. W. Sanford, S. Tapia, F. J. Vrba, *ApJ* **237** (1980) 154.
Spectroscopy: J. Thorstensen *et al.*; P. A. Charles *et al.*
X-ray observations: P. A. Charles *et al.*
Duplicity: Light variations with period 1.28 ... 1.31^d due to orbital motion are suspected (V. M. Lyutiy, S. Yu. Shugarov, *Sov Astr Lett* **5** (1979) 383).
Identification: from finding chart by Thorstensen *et al.*
Classification: X-ray nova.

V1370 Aql — N

(N Aql 1982)

Discovered by M. Honda, Japan, 1982 January 27 (*IAU Circ* 3661).

Position:	19 20 50.08	+ 02 23 34.8	(POSS)
	19 20 50.14	+ 02 23 35.4	(2 outburst observations)
	38.813	− 5.947	(G.C.)

Range: 6: – 19.5p LCT: Cb t_3: ?

Finding chart: L. Rosino, T. Iijima, S. Ortolani, *MN* **205** (1983) 1069: H. W. Duerbeck, M. Geffert, *IBVS* 2260 (1983); G. Hacke, *MVS* **9** (1983) 147.

Light curve: L. Rosino, T. Iijima, S. Ortolani, *MN* **205** (1983) 1069; G. Hacke, *MVS* **9** (1983) 147.

Spectroscopy: L. Rosino, T. Iijima, S. Ortolani, *MN* **205** (1983) 1069 – trac, ident, rv; A. Okazaki, A. Yamasaki, *ApSS* **119** (9186) 89 – trac, ident, rv; M. A. J. Snijders, T. J. Batt, M. J. Seaton, J. C. Blades, D. C. Morton, *MN* **211** (1984) 7p – trac, ident; Y. Andrillat, *MN* **203** (1983) 5p – trac, [Ne III] lines.

UV observations: M. A. J. Snijders, T. J. Batt, M. J. Seaton, J. C. Blades, D. C. Morton, *MN* **211** (1984) 7p.

IR observations: M. F. Bode, A. Evans, D. C. B. Whittet, D. K. Aitken, P. F. Roche, B. Whitmore, *MN* **207** (1984) 897; R. D. Gehrz, E. P. Ney, G. L. Grasdalen, J. A. Hackwell, H. A. Thronson, Jr., *ApJ* **281** (1984) 303; P. F. Roche, D. K. Aitken, B. Whitmore, *MN* **211** (1984) 535; P. M. Williams, A. J. Longmore, *MN* **207** (1984) 139; R. M. Catchpole, I. S. Glass, G. Roberts, J. Spencer Jones, P. Whitelock, *SAAO Circ* **9** (1985) 1.

Identification: from published finding charts.

Classification: nova with dust formation; the light curve depression is only 1^m.

N Aql 1984 N

Discovered by M. Honda, Japan, on a photograph taken 1984 December 2, as a 10^m star. The nova is not visible on a photograph of 1984 November 29 (*IAU Circ* 4020).

Position:	19 14 05.73	+ 03 37 55.8	(POSS; diffuse object at plate limit)
	19 14 05.79	+ 03 37 55.5	(GPO plate, May 1986)
	19 14 05.82	+ 03 37 55.4	(H. Kosai, *IAU Circ* 4023 (1984))
	39.124	− 3.880	(G.C.)

Range: **10**p – 21p LCT: ? t_3: ?

Spectroscopy: H. Kosai, *IAU Circ* 4023 (1984) – descr.

Identification: from published precise position and GPO plate.

Classification: poorly known nova.

KY Ara **UG?**

(HV 7987)

Discovered by C. D. Boyd on Harvard plates. Maximum light occurred in 1937 July. Only 4 positive observations (H. Shapley, E. H. Boyce, C. D. Boyd, *HA* **90** (1939) 244).

Position:	18 04 11.96	– 54 56 40.4	(SRC)
	18 04 18	– 54 56 54	(H. Shapley, E. H. Boyce, C. D. Boyd, *HA* **90** (1939)
	338.908	– 16.121	(G.C.)

Range: 15.1p – 18p? LCT: ? t_3: ?

Finding chart: S. Wyckoff, P. A. Wehinger (1978); V. P. Tsesevich, M. S. Kazanasmas (1971).

Light curve: H. Shapley, E. H. Boyce, C. D. Boyd, *HA* **90** (1939) 244.

Identification: from Harvard plate MF 23447, taken 1937 July 4/5. The object is not clearly recognized; the candidate from the finding charts is marked. Spectroscopic observations are necessary in order to check the identification.

Classification: sharp maximum, possibly small outburst amplitude, no spectroscopic observations during outburst are available. The GCVS4 classification is UG.

OY Ara **NA**

(N Ara 1910, 98.1910 Ara, HV 3305, HD 149990)

Discovered by W. Fleming on a Harvard plate taken 1910 April 4, as a star of $6^{m}_{.}0$. Invisible until 1910 March 19 ([12^{m}) (*AN* **186** (1910) 95).

Position:	16 36 55.275	– 52 20 04.0	(SRC)
	16 36 55	– 52 19 35	(E.C. Pickering, *AN* **186** (1910) 95)
	333.902	– 3.937	(G.C.)

Range: 6.0p (5.1) – 17.5p LCT: Bb? t_3: 80^{d}

Light curve: A. D. Walker, *HA* **84** (1923) 189.

Spectroscopy: A. J. Cannon, *HA* **76** (1916) 19 – descr.

Identification: from Harvard plate A 10364, taken 1911 March 29; the identification on the atlas plate is not unambiguous since the marked object is surrounded by several stars within a few arc seconds.

Classification: moderately fast nova with fairly well covered light curve, which shows peculiar secondary maximum.

Classification: The GCVS4 classifies the object as N:. No spectroscopic outburst data are available; its reality is not established.

AB Boo N??

(N Boo 1877, N Com 1877, Zi 1047, BD +21°2606)

Discovered by F. Schwab, Ilmenau, who observed the object as a 5^m star from 1877 May 30 to 1877 July 14. Its position nearly coincides with BD +21°2606 ($9^m.4$) (*AN* **155** (1901) 220, *AN* **156** (1901) 349).

Position:			
(1)	14 04 43.20	+20 58 55.0	(POSS; BD +21°2606)
(2)	14 04 45.99	+20 58 28.15	(POSS)
(3)	14 04 42.84	+20 59 28.4	(POSS)
(4)	14 04 43.33	+21 01 07.5	(POSS)
(5)	14 04 43.45	+20 56 31.0	(POSS)
	16.706	+71.604	(G.C.)

Range: 4.5v – ? LCT: ? t_3: ?

Finding chart: A. Sh. Khatisov (1971).
Identification: Khatisov identifies BD +21°2606 with Schwab's object. This identification is based on Schwab's crude estimate of the position and may be incorrect. We have measured four additional faint candidates in the field of BD +21°2606.
Classification: The GCVS4 classification is N:. Its existence is not well established.

N Boo 1962 N?

(S 10808)

Discovered by H. Gessner on a Sonneberg plate taken 1962 December 3 (*IBVS* 1428 (1978)).

Position:			
	14 40 56.18	+13 54 31.9	(POSS)
	14 40 54	+13 56	(H. Gessner, *IBVS* 1428 (1978))
	11.320	+60.650	(G.C.)

Range: 10.5p – 20p? LCT: ? t_3: ?

Finding chart: H. Gessner, *MVS* **8** (1978) 66.
Identification: despite the remark by Gessner (*IBVS* 1428 (1978)), that no counterpart is found on the POSS print, there is a faint blue star near the

position of Gessner's finding chart. Its position is given above. Spectroscopic verification is necessary.

Classification: existence of S 10808 based on only one photographic plate; confirmation required.

CG CMa N?

Discovered by J. Verlooy on Franklin-Adams plates taken in January 1934. Maximum light was reached on 1934 January 12 (A. van Hoof, *Louvain Publ* **12** (1948) 13).

Position:	07 01 59.55	– 23 41 03.2	(SRC)
	07 02 00	– 23 41 02	(A. van Hoof, *Louvain Publ* **12** (1948) 13)
	235.596	– 7.978	(G.C.)

Range: 13.7p – 15.9p? LCT: A? t_3: –

Finding chart: A. van Hoof, *Louvain Publ* **12** (1948) 13.
Light curve: A. van Hoof, *Louvain Publ* **12** (1948) 13.
Identification: from Harvard plates RB 4799, taken 1934 January 17, and RB 4827, taken 1934 January 22.
Classification: The GCVS classification is N:. No spectroscopic information is available. The possibly low outburst amplitude makes DN variability more likely. If CG CMa is indeed a nova, it lies at a very large distance in the direction of the galactic anticenter.

RS Car N

(N Car 1895, HV 47, HD 96830)

Discovered by W. Fleming on Harvard objective prism plates. It is visible on plates taken between 1895 April 8 and November 13 (E. C. Pickering, *AN* **139** (1895) 119).

Position:	11 06 01.46	– 61 39 49.2	(SRC; empty field; estimated position)
	11 05 58	– 61 40 13	(E. C. Pickering, *AN* **139** (1895) 119)
	291.098	– 1.469	(G.C.)

Range: 7.0p (5.0) – [22j LCT: B? t_3: ?

Finding chart: A. J. Cannon, M. W. Mayall, *HA* **112** (1949) 166.
Light curve: A. D. Walker, *HA* **84** (1923) 189; G. Cecchini, L. Gratton (1941) 50.
Spectroscopy: A. J. Cannon, *HA* **76** (1916) 19 – descr.

Identification: from Harvard plate B 12988, taken 1895 April 9.
Classification: nova; the fragmentary light curve shows fluctuations.

V351 Car — M

(BV 1543)

Discovered by R. Knigge on sky patrol plates of the Boyden Observatory (*IBVS* 765 (1983)), and classified as nova. D. J. MacConnell (*IBVS* 799 (1973)) finds that the star is a long period variable of spectral type M.

Position:	10 43 51.41	− 71 48 06.6	(SRC)
	10 43 52	− 71 48 04	(R. Knigge, *IBVS* 765 (1973))
	293.472	− 11.550	(G.C.)

Range: 12.5 p − ? LCT: M t_3: −

Finding chart: R. Knigge, *IBVS* 765 (1973).
Light curve: R. Knigge, *IBVS* 765 (1973).
Spectroscopy: D. J. MacConnell, *IBVS* 799 (1973) − descr.
Identification: from R. Knigge's finding chart.
Classification: long period variable.

V365 Car — NB

(He 3-558)

Discovered by K. G. Henize as Hα emission-line object (*ApJ Suppl* **14** (1967) 125), classified by W. Liller as slow nova. The object became visible on 1948 July 2, maximum occurred 1948 September 25.

Position:	11 01 09.30	− 58 11 17.2	(SRC)
	11 01 12	− 58 11	(K. G. Henize, W. Liller, *ApJ* **200** (1975) 694)
	289.168	+ 1.493	(G.C.)

Range: **10.1**p − 21.6j LCT: D t_3: 530^d

Finding chart: K. G. Henize, W. Liller, *ApJ* **200** (1975) 694.
Light curve: K. G. Henize, W. Liller, *ApJ* **200** (1975) 694.
Spectroscopy: K. G. Henize, W. Liller, *ApJ* **200** (1975) 694.
Identification: from Henize and Liller's finding chart. The authors state that the nova lies near the galactic cluster NGC 3532, but membership can be excluded.
Classification: very slow nova with well-documented light curve.

N Car 1953 N

(NSV 04884)

Discovered by L. Perek as emission-line object on objective prism plates of the Tonantzintla Schmidt telescope, taken 1953 February 8/9 and 1953 March 10/11 (*BAC* **11** (1960) 256).

Position:	10 29 29.37	− 59 42 59.35	(ESO)
	10 29 31	− 59 43 24	(L. Perek, *BAC* **11** (1960) 356)
	286.232	− 1.726	(G.C.)

Range: 14.5p − 19p LCT: ? t_3: ?

Light curve: H. W. Duerbeck, *IBVS* 2502 (1984).
Spectroscopy: L. Perek, *BAC* **11** (1960) 256 − descr.
Identification: from Harvard plates B 76601, taken 1953 January 23/24, and B 76679, taken 1953 March 19/20.
Classification: nova, poorly known.

N Car 1971 N

Discovered by D. J. MacConnell, E. Prato, and C. Briceño on an objective prism plate taken with the Curtis Schmidt telescope, Cerro Tololo, 1971 February 18 (*IBVS* 1476 (1978)).

Position:	10 37 59.08	− 62 58 27.3	(SRC)
	10 37 59	− 62 58 33	(D. J. MacConnell, E. Prato, C. Briceño, *IBVS* 1476 (1978))
	288.732	− 4.503	(G.C.)

Range: 13? − 18? LCT: ? t_3: ?

Finding chart: D. J. MacConnell, E. Prato, C. Briceño, *IBVS* 1476 (1978).
Spectroscopy: D. J. MacConnell, C. Prato, E. Briceño, *IBVS* 1476 (1978) − strong Hα emission on the discovery plate, not seen on a plate taken 13 months later.
Identification: from the position and finding chart of MacConnell *et al.*, not verified by additional observations.
Classification: poorly known nova.

N Car 1972 — N

Discovered by D. J. MacConnell, E. Prato, and C. Briceño on an objective prism plate taken with the Curtis Schmidt telescope, Cerro Tololo, 1972 March 22 (*IBVS* 1476 (1978)).

Position: 10 36 35.04 – 62 52 51.6 (SRC)
10 36 35 – 62 52 52 (D. J. MacConnell, E. Prato, C. Briceño, *IBVS* 1476 (1978))
288.547 – 4.048 (G.C.)

Range: 13? – 18? LCT: ? t_3: ?

Finding chart: D. J. MacConnell, E. Prato, C. Briceño, *IBVS* 1476 (1978).
Spectroscopy: D. J. MacConnell, C. Prato, E. Briceño, *IBVS* 1476 (1978) – strong Hα emission on the discovery plate, not seen on a plate taken 13 months earlier.
Identification: from the position and finding chart of MacConnell *et al.*, not verified by additional observations. Blend of several stars.
Classification: poorly known nova.

η Car — SDOR

(η Argus, CPD –59°2620, CoD –59°3306, HD 93308, HR 4210, Boss 2871, MWC 214, GC 14799, N Car 1843)

Peculiar high-luminosity object whose spectrum at times resembled that of a nova. The variability was discovered by Burchell (1827).

Position: 10 43 06.87 – 59 25 16.5 (Perth 70 catalogue)
287.597 – 0.630 (G.C.)

Range: **–0.8**v – 9.0v LCT: C? E? t_3: 16.7 years

Finding chart: A. J. Cannon, M. W. Mayall, *HA* **112** (1949) 148; G. de Vaucouleurs, O. J. Eggen, *PASP* **64** (1952) 189.
Light curve: R. T. A. Innes, *Cape Ann* **9** (1903) 78B; D. J. K. O'Connell, *Vistas* **2** (1956) 1165; A. Feinstein, H. G. Marraco, *AsAp* **30** (1974) 271.
Spectroscopy: A. LeSueur, *Proc R Soc* **18** (1970) 245, **19** (1870) 18; P. W. Merrill, *ApJ* **67** (1928) 391 – ident; H. Spencer Jones, *MN* **91** (1931) 794 – ident; A. D. Thackeray, *MN* **113** (1953) 211 – phot, ident; E. Gaviola, *ApJ* **118** (1953) 234 – phot, ident; A. D. Thackeray, *MN* **124** (1962) 251 – IR spectra; N. R. Walborn, M. H. Liller, *ApJ* **211** (1977) 181 – phot (also historical spectra).

Nebular remnant: E. Gaviola, *ApJ* **111** (1949) 408; A. D. Thackeray, *MN* **110** (1950) 524.

Classification: The GCVS4 classification is SDOR. Despite the fact that luminosity and time scale are different from those of novae, the 'rapid' decline around 1860 and the recovery around 1950 are reminiscent of novae with light curve type C and dust formation. A review is given by A. M. van Genderen and P. S. Thé, *Space Sci Rev* **39** (1984) 317.

BC Cas **NA:**

(N Cas 1929, 345.1931 Cas, SVS 254)

Discovered by S. J. Belyavsky on Simeis plates and classified originally as long period variable (*AN* **243** (1931) 115).

Position:	23 48 48.71	+ 60 01 29.2	(POSS)
	23 48 51	+ 60 01 24	(S. J. Belyavsky, *AN* **243** (1931) 115)
	115.544	− 1.703	(G.C.)

Range: 10.7p − 17.4p LCT: B? t_3: 75^d

Finding chart: N. B. Perova, A. S. Sharov, *PZv* **11** (1956) 388; Yu. N. Efremov (1961).

Light curve: H. W. Duerbeck, *IBVS* 2490 (1984).

Identification: Efremov's identification is confirmed through Harvard plate I 47668, taken 1929 November 5/6.

Classification: poorly known object with small outburst amplitude. No spectroscopic information is available. Moderately fast nova?

V630 Cas **UG:**

(OV 29)

Discovered by B. S. Whitney, Norman, Oklahoma, on plates taken in 1950. Maximum occurred around 1950 November 14 (*IBVS* 797 (1973)).

Position:	23 46 22.67	+ 51 10 59.0	(POSS)
	23 46 30	+ 51 11	(B. S. Whitney, *IBVS* 797 (1973))
	113.103	− 10.213	(G.C.)

Range: **12.3**p − 17.1p LCT: DN? t_3: $\sim 50^d$

Finding chart: B. S. Whitney, *IBVS* 797 (1973).

Light curve: B. S. Whitney, *IBVS* 797 (1973).

Identification: from Whitney's finding chart, confirmed by spectroscopic observations (H. W. Duerbeck, unpublished)

PURCHASE ORDER # 2547661 **DATE** 08/11/87

A REFERENCE CATALOGUE AND ATLAS OF GALACTIC NOVAE. Duerbeck, Hilmar W., 1948-. Boston : D. Reidel, 1987.

;

astr ; EN/AJ; T;

RETURN THIS COPY WITH MATERIAL OR REPORT

ISBN/ISSN	EST. PRICE PER COPY	FUND	VENDOR	RL
	$45.01	astrm	bh	ACQ

Classification: B. S. Whitney (*IBVS* 797 (1973)) notes the blue colour at maximum light, which is atypical for novae. The maximum was followed by a rapid decline. V630 Cas is likely to be a WZ Sge-type dwarf nova with rare outbursts.

MT Cen **N:**

(N Cen 1931)

Discovered by J. Uitterdijk on Johannesburg plates taken in 1931 May. The object appears first on plates of 1931 May 9, maximum is reached on May 10 (*BAN* **7** (1934) 176).

Position:			
	11 41 35.88	− 60 16 59.9	(SRC)
	11 41 36	− 60 16 54	(J. Uitterdijk, *BAN* **7** (1934) 176)
	294.736	+ 1.234	(G.C.)

Range: **8.35**p – 22j LCT: Cb? t_3: ?

Finding chart: J. Uitterdijk, *BAN* **7** (1934) 176.
Light curve: J. Uitterdijk, *BAN* **7** (1934) 176; H. W. Duerbeck, *IBVS* 2490 (1984).
Identification: from Harvard plates MF 15521, taken 1931 May 23/24, and MF 15676, taken 1931 July 7/8. The exnova is near the limit of the SRC plate.
Classification: very fast nova of light curve type Cb (or A, if one point of the light curve is discarded). No spectroscopic confirmation is available.

V359 Cen **N:**

(vG 27)

Discovered by A. Opolski on Franklin-Adams plates. The object is visible on 19 plates taken between 1939 April 20 and 27; the maximum was probably missed (*Lwów Contr* **4** (1935)).

Position:			
	11 55 42.39	− 41 29 26.0	(SRC)
	11 55 54	− 41 29	(A. Opolski, *Lwów Contr* **4** (1935))
	292.414	+ 20.014	(G.C.)

Range: 13.8p – 21.0j LCT: ? t_3: ?

Finding chart: A. Opolski, *Lwów Contr* **4** (1935).

Light curve:	A. Opolski, *Lwów Contr* **4** (1935).
Identification:	from Harvard plates RB 914, taken 1930 April 22/23, and RB 932, taken 1930 April 25/26, and Opolski's finding chart.
Classification:	poorly known object without spectroscopic confirmation. It cannot be decided whether it is a nova or a dwarf nova with rare outbursts.

V812 Cen — N

(N Cen 1973)

Discovered by D. J. MacConnell, E. Prato and C. Briceño on objective prism plates taken with the Curtis Schmidt telescope, Cerro Tololo, 1973 April 23 and 25. The spectrum shows the diffuse enhanced stage of the nova (*IBVS* 1476 (1978)).

Position:	13 10 47.88	− 57 24 52.05	(ESO QB survey; decline)
	13 10 47.79	− 57 24 52.2	(SRC)
	13 10 47.9	− 57 24 50	(D. J. MacConnell, E. Prato, C. Briceño, *IBVS* 1476 (1978))
	305.943	+ 5.061	(G.C.)

Range: 11p (9) – 18j LCT: ? t_3: ?

Finding chart:	D. J. MacConnell, E. Prato, C. Briceño, *IBVS* 1476 (1978).
Spectroscopy:	D. J. MacConnell, E. Prato, C. Briceño, *IBVS* 1476 (1978).
Identification:	from ESO Schmidt plate QB 173, taken 1974 February 25.
Classification:	nova, poorly known.

V822 Cen — XND

(Cen X-4)

Optical counterpart of the recurrent X-ray transient source Cen X-4 (C. R. Canizares, J. E. McClintock, J. E. Grindlay, *ApJ* **236** (1980) L55).

Position:	14 55 19.52	− 31 28 08.8	(SRC)
	14 55 19.63	− 31 28 09.0	(C. R. Canizares, J. E. McClintock, J. E. Grindlay, *ApJ* **236** (1980) L55)
	332.242	+ 23.881	(G.C.)

Range: **12.8**p – 19p LCT: ? t_3: 20^d

Finding chart:	C. R. Canizares, J. E. McClintock, J. E. Grindlay, *ApJ* **236** (1980) L55.

Light curve: C. R. Canizares, J. E. McClintock, J. E. Grindlay, *ApJ* **236** (1980) L55.
Spectroscopy: C. R. Canizares, J. E. McClintock, J. E. Grindlay, *ApJ* **236** (1980) L55.
X-ray observations: C. R. Canizares, J. E. McClintock, J. E. Grindlay, *ApJ* **236** (1980) L55; L. J. Kaluzienski, S. S. Holt, J. H. Swank, *ApJ* **241** (1980) 779.
Duplicity: $0^m.2$ modulation with P = 0.629075 d (S. A. Ilovasiky, C. Chevalier, M. van der Klis, J. van Paradijs, H. Pedersen, *IAU Circ* 4264 (1986)).
Identification: from the finding chart of C. R. Canizares *et al.*
Classification: X-ray nova.

N Cen 1986 No. 1 — ZAND

(Liller's novalike object in Cen)

Discovered by W. Liller, Viña del Mar, Chile, 1986 January 3, as a star of $10^m.9$. Maximum was reached around 1986 January 13 (*IAU Circ* 4180).

Position:	13 17 42.36	− 55 34 30.1	(SRC)
	13 17 40	− 55 34 54	(W. Liller, *IAU Circ* 4180 (1986))
	307.079	+ 6.798	(G.C.)

Range: **7.5**v − 14.5v LCT: ? t_3: 121^d

Finding chart: R. H. McNaught, G. Dawes, *IBVS* 2927 (1986).
Light curve: W. Liller, *IAU Circ* 4180 (1986); R. H. McNaught, G. Dawes, *IBVS* 2927 (1986).
Spectroscopy: H. W. Duerbeck, W. C. Seitter, *MN* (in press) – trac, descr.
Identification: from McNaught and Dawes' finding chart.
Classification: spectroscopic observations at minimum show a symbiotic spectrum with strong stellar continuum.

N Cen 1986 No. 2 — N

Discovered by R. H. McNaught, Siding Spring Observatory, 1986 November 22, as a star of $5^m.6$. It was [$7^m.2$ on November 10 (*IAU Circ* 4274).

Position:	14 32 13.33	− 57 24 30.6	(SRC)
	14 32 13.47	− 57 24 31.1	(R. H. McNaught, *IAU Circ* 4274 (1986))
	316.574	+ 2.453	(G.C.)

Range: 4.6v − 18.6 ... 20.3j LCT: Cb t_3: ?

Spectroscopy: L. H. Barrera, G. Le-Cerf, N. Vogt, *IAU Circ* 4299 (1987) – descr.
Identification: from the published precise position. Finding chart in Appendix.
Classification: slow nova.

IV Cep NA

(N Cep 1971)

Discovered by Y. Kuwano, Japan, 1971 July 10.5, as a star of 8^m. Maximum light occurred on July 7.9 (7^m) (*IAU Circ* 2340).

Position: 22 02 46.83 + 53 15 48.0 (7 outburst observations)
99.614 – 1.638 (G.C.)

Range: 7.5B (7.0) – 17.1B LCT: Ba t_3: 37^d

Finding chart: G. K. Walker, M. J. Keyes, *IBVS* 583 (1971); H. Kosai, *Tokyo Bull 2nd Ser* **214** (1971) 2515; L. Kohoutek, P. Klawitter, *AsAp Suppl* **11** (1973) 347.
Light curve: L. Kohoutek, P. Klawitter, *AsAp Suppl* **11** (1973) 347 (also pre-outburst observations); A. P. Cowley, D. J. MacConnell, J. Toney, *PASP* **85** (1973) 309; G. C. Aikman, R. W. Hilditch, F. Younger, *PASP* **85** (1973) 756; S. Sato, T. Maihara, H. Okuda, *PASJ* **25** (1973) 571; R. Burchi, V. d'Ambrosio, *IBVS* 2703 (1985).
Spectroscopy: J. C. Thomas, A. P. Cowley, D. J. MacConnell, J. Toney, *PASP* **85** (1973) 347 – trac, rv; G. C. Aikman, R. W. Hilditch, F. Younger, *PASP* **85** (1973) 756 – trac, rv; J. D. R. Bahng, *MN* **158** (1972) 151 – trac, spectrophotometry; C. Fehrenbach, Y. Andrillat, *CR Sér B* **273** (1971) 572, **274** (1972) 1179 – phot, rv.
IR observations: S. Sato, T. Maihara, H. Okuda, *PASJ* **25** (1973) 571.
Identification: from published finding charts.
Classification: well observed fast nova.

WX Cet UGWZ

(N Cet 1963, BV 416)

Discovered by W. Strohmeier on Bamberg plates taken in 1963 September-October (*IBVS* 47 (1964)).

Position: 01 14 37.52 – 18 12 10.4 (POSS)
01 14 37.6 – 18 12 10.7 (S. Wyckoff, P. A. Wehinger (1978))
157.007 – 79.076 (G.C.)

Range: 10.5p – 18.0p LCT: DN t_3: ?

Finding chart: S. Wyckoff, P. A. Wehinger (1978).
Light curve: negative search for outbursts in 1928 – 1976: I. Meinunger, *MVS* **7** (1976) 192; positive search by S. Gaposchkin, *IBVS* 1204 (1976), maxima of unequal height in 1938 (9.4), 1939 (10.2), 1945 (13.5), and 1963 (10.5).
Spectroscopy: R. A. Downes, B. Margon, *MN* **197** (1981) 35p – minimum spectra.
Identification: from Wyckoff and Wehinger's finding chart.
Classification: S. Gaposchkin (*IBVS* 1204 (1976)) classifies the star as UG variable or 'subnova'. R. A. Downes and B. Margon (*MN* **197** (1981) 35p) state that it is similar to dwarf novae of WZ Sge type.

RR Cha NA:

(N Cha 1953, S 5346)

Discovered by C. Hoffmeister, Sonneberg, on sky patrol plates. The nova is seen between 1953 April 8 and 1953 July 13; maximum light was missed (*IAU Circ* 1671 (1959)).

Position: 13 20 55.70 – 82 04 07.4 (SRC)
13 20 56.54 – 82 04 06.1 (C. Hoffmeister, *Sterne* **36** (1960) 93)
304.165 – 19.541 (G.C.)

Range: 7.1p – 19.3j LCT: B? t_3: 60^d

Finding chart: C. Hoffmeister, *Sterne* **36** (1960) 93.
Light curve: C. Hoffmeister, *AN* **285** (1959) 149.
Identification: from Hoffmeister's finding chart.
Classification: nova without spectroscopic confirmation, but with typical moderately fast light curve development and large amplitude.

X Cir NB

(N Cir 1926, N Cir No. 2)

Discovered by F. Becker on an objective prism plate taken at La Paz 1927 May 21 (*AN* **237** (1927) 71).

Position: 14 38 33.64 – 64 58 50.0 (SRC; empty field or object at plate limit)
14 38 36 – 65 00 (F. Becker, *AN* **237** (1927) 71)
314.259 – 4.797 (G.C.)

Range: 6.5p (6.2) – [23j LCT: ? t_3: 170^d

Light curve: A. J. Cannon, *HB* 872 (1930) 1; G. Cecchini, L. Gratton (1941) 135.
Spectroscopy: F. Becker, *AN* **237** (1927) 71.
Identification: from Harvard plates MF 11405, MF 11412, MF 11463 and MF 11540, taken between 1928 February 28 and April 19, all poorly guided.
Classification: slow nova; the maximum is nearly covered, subsequent stages are not observed.

AI Cir N:

(N Cir 1914, N Cir No. 3, HV 5121)

Discovered by E. B. Florence on Harvard plates. The nova was [$10^m.9$ on 1914 June 3, $10^m.9$ on 1914 June 13 (*HB* 883 (1931) 23).

Position:	14 45 03.02	– 68 39 09.2	(SRC; extremely faint)
	14 45 04	– 68 39 06	(H. H. Swope, *HB* 883 (1933) 23)
	313.280	– 8.395	(G.C.)

Range: 10.9p – 23j LCT: ? t_3: ?

Finding chart: H. Shapley, H. H. Swope, *HB* 855 (1931) 13.
Light curve: H. H. Swope, *HB* 883 (1931) 23.
Identification: from Harvard plates B 44496, 1914 May 29, B 44584, 1914 June 13, and B 44645, 1914 June 23. On the SRC plate is an extremely faint object, essentially only a grain fluctuation.
Classification: only six positive observations indicate a light curve with 1^m oscillation. No spectroscopic information is available, but the amplitude indicates nova type.

AR Cir N?

(N Cir 1906, N Cir No. 1, 102.1907 Cir, HV 2954)

Discovered by H. Leavitt on Harvard plates. The object is brighter than 12^m on all plates taken in 1906 (February 14 to August 24). On earlier plates, the star is [$13^m.0$ (*HC* 130 (1907), *AN* **175** (1907) 333).

Position:	14 44 19.28	– 59 47 56.6	(SRC)
	14 44 19	– 59 47 42	(E. C. Pickering, *HC* 130 (1907))
	317.039	– 0.372	(G.C.)

Range: 10.3p – 15p LCT: D or E t_3: 415^d

Light curve: A. D. Walker, *HA* **84** (1923) 189; C. Payne-Gaposchkin (1957) 16; G. Cecchini, L. Gratton (1941) 83, 84.
Identification: from Harvard plates A 7757, taken 1906 June 27, and A 8252, taken 1907 April 21. The nova coincides with a star of 15^m on the SRC plate. It is probably an unresolved companion of late type (H. W. Duerbeck, W. C. Seitter, *ApSS* **131** (1987) 467.
Classification: very slow nova without spectroscopic confirmation and probably small amplitude.

AL Com UGSS

(N Com 1961, Rosino's object)

Discovered by L. Rosino on Asiago plates taken 1961 November 17 (*IAU Circ* 1782).

Position:	12 29 54.47	+ 14 37 15.85	(POSS)
	12 29 54	+ 14 37 18	(GCVS4)
	282.869	+ 76.471	(G.C.)

Range: **13.0**p – 20.0p LCT: DN, pec t_3: 30^d

Finding chart: F. Bertola, *AAp* **27** (1964) 298; N. Vogt, F. Bateson (1982).
Light curve: F. Bertola, *AAp* **27** (1964) 298.
Identification: from Bertola's finding chart.
Classification: dwarf nova with mean cycle length of 325^d. AL Com is in the field of the galaxy NGC 4501 (M 88). Its light curve shows a deep minimum of $\sim 1^d$ duration during the later stages of the outburst.

V394 CrA N

(N CrA 1949, N Sco 1949)

Discovered by L. E. Erro, Tonantzintla Observatory, 1949 March 23, as a star of $7^m.5$. It was [$12^m.5$ the night before (*IAU Circ* 1208).

Position:	(1) 17 56 58.165	– 39 00 29.4	(SRC, star 19^m)
	(2) 17 56 58.00	– 39 00 26.6	(SRC, star 20^m)
	(3) 17 56 58.31	– 39 00 25.15	(SRC, star 21^m)
	17 56 58	– 39 00 25	(W. H. van den Bos, *MNASSA* **8** (1949) 35)
	352.836	– 7.715	(G.C.)

Range: **7.5**p – 20:j LCT: ? t_3: ?

Spectroscopy: L. E. Erro, *IAU Circ* 1211 (1949).
Identification: from Harvard plates MF 38199, B 74552, B 74684 and B 74786, taken between 1949 March 25/26 and May 26/27. The nova coincides with a group of three stars, which are listed above.
Classification: poorly known fast nova.

V655 CrA **N**

(N CrA 1967)

Discovered by N. Sanduleak on objective prism plates taken with the Curtis Schmidt telescope, Cerro Tololo, 1967 June 30 and July 30 (*IBVS* 368 (1969)).

Position:			
(1)	18 21 20.80	– 37 01 23.9	(SRC)
(2)	18 21 20.52	– 37 01 25.95	(SRC)
	18 21 24	– 37 02	(N. Sanduleak, *IBVS* 368 (1969))
	356.859	– 11.057	(G.C.)

Range: 8p – 17.6j LCT: ? t_3: ?

Finding chart: N. Sanduleak, *IBVS* 368 (1969).
Spectroscopy: N. Sanduleak, *IBVS* 368 (1969) – descr.
Identification: from Sanduleak's finding chart. The nova is an unresolved double star with components of equal brightness.
Classification: poorly known nova.

V693 CrA **NA**

(N CrA 1981)

Discovered by M. Honda, Japan, 1981 April 2.8, as a star of $7^{m}_{.}0$. It was not visible on 1981 April 1 (*IAU Circ* 3590).

Position:		
18 38 33.6	– 37 34 09	(A. C. Gilmore, *IAU Circ* 3591 (1981))
357.830	– 14.392	(G.C.)

Range: **7.0**v – 23j LCT: A t_3: 12^d

Light curve: N. Brosch, *IBVS* 2003 (1981); J. A. R. Caldwell, *IBVS* 2417 (1981); W. S. G. Walker, B. F. Marino, *NZAS Publ* **10** (1982) 48; B. S. Shylaya, *ApSS* **104** (1984) 163.

Spectroscopy: N. Brosch, *AsAp* **107** (1982) 300 – trac, ident; B. S. Shylaya, *ApSS* **104** (1984) 163 – trac.
UV observations: R. E. Williams, E. P. Ney, W. M. Sparks, S. C. Starrfield, S. Wyckoff, J. W. Truran, *MN* **212** (1985) 753.
IR observations: R.M. Catchpole, I. S. Glass, G. Roberts, J. Spencer Jones, P. Whitelock, *SAAO Circ* **9** (1985) 1.
Identification: from Harvard plates DSB 788, taken 1981 May 4, and DSB 777, taken 1981 August 27; extremely faint star on SRC plate.
Classification: very fast nova.

T CrB NR

(N CrB 1866, BD +26°2765, HD 143454, GC 21491, HR 5958, MWC 413)

Discovered by J. Birmingham, England, 1866 May 12, as a 2^{m} star. A second outburst was observed on 1946 May 12 (*AN* **67** (1866) 197).

Position:			
	15 57 24.50	+26 03 38.25	(3 recent observations)
	15 57 24.80	+26 03 36.7	(8 outburst observations)
	42.373	+48.165	(G.C.)

Range: **2.0**p – 11.3p LCT: Ar t_3: 6.8^{d}

Finding chart: M. Humason (1938); A. Sh. Khatisov (1971); G. Williams (1983).
Light curve: J. N. Lockyer, *Phil Trans* **182A** (1891) 297; F. W. Wright, *HB* 918 (1946) 9; E. Pettit, *PASP* **58** (1946) 152, 213, 255; C. Payne-Gaposchkin, F. W. Wright, *ApJ* **104** (1946) 75; W. W. Morgan, A. J. Deutsch, *ApJ* **106** (1947) 362; K. C. Gordon, G. E. Kron, *IBVS* 1610 (1979); C. Payne-Gaposchkin (1957) 104; G. Cecchini, L. Gratton (1941) 31, 32; S. J. Kenyon (1986) 92.
Spectroscopy: J. N. Lockyer, *Phil Trans* **182A** (1891) 397 – descr; W. Huggins, *AN* **67** (1866) 29 – descr; W. W. Morgan, A. J. Deutsch, *ApJ* **106** (1947) 362 – phot, ident; G. H. Herbig, F. J. Neugebauer, *PASP* **58** (1946) 196 – phot, ident, rv; D. B. McLaughlin, *PASP* **58** (1946) 159 – descr, comparison with 1866 outburst; R. F. Sanford, *PASP* **58** (1946) 156 – descr, rv; R. F. Sanford, *PASP* **59** (1947) 87 – phot, rv, orbital motion; M. Bloch, J. Dufay, Ch. Fehrenbach, Tcheng M. L., *AAp* **9** (1946) 157 – phot, trac, ident.
G. Williams (1983) – minimum spectrum, trac; R. P. Kraft, *ApJ* **139** (1964) 457 – minimum spectrum, phot.
UV observations: A. Cassatella, P. Patriachi, P. L. Selvelli, L. Bianchi, C. Cacciari, A. Heck, M. Perryman, W. Wamsteker, in Advances in Ultraviolet Astronomy, *NASA CP-2238* (1982), eds. Y. Kondo, J. M. Mead, R. D. Chapman, p. 482.

Duplicity:	spectroscopic binary, P = 227.3^d (R. P. Kraft, *ApJ* **127** (1958) 620; S. J. Kenyon, M. R. Garcia, *AJ* **91** (1986) 125); visual binary, component B = 12^m, d = 0″.2 (?)).
Nova shell:	R. E. Williams, *IAU Coll. 42 = Bamb Ver* **9**, 121 (1977) 242.
Identification:	from published finding charts.
Classification:	recurrent nova; outbursts in 1866 and 1946.

AP Cru N:

(N Cru 1935, 615.1936 Cru, P 3555)

Discovered as variable by D. J. K. O'Connell, Riverview College Observatory, later classified as nova (*AN* **259** (1936) 399).

Position:	12 28 28.59	− 64 09 50.8	(SRC)
	12 28 29	− 64 10	(D. J. K. O'Connell, *AN* **259** (1936) 399)
	300.765	− 1.653	(G.C.)

Range: 10.7p (9.0) – 21.7j LCT: ? t_3: ?

Light curve:	D. J. K. O'Connell, *Riv Publ* **2** (1948) 68.
Identifications:	from Harvard plates B 59672, taken 1935 April 30/May 1, B 59758, taken 1935 May 29/30, and B 59984, taken 1935 June 30/July 1.
Classification:	poorly known object; amplitude suggests nova. No spectroscopic information is available. P. R. Amnuel and O. H. Guseinov suspect coincidence with the X-ray source 1225-64 (*PZv* **19** (1973) 19).

V404 Cyg NA

(N Cyg 1938, 100.1938 Cyg)

Discovered by A. A. Wachmann, Hamburger Sternwarte, after maximum light, which must have occurred between 1938 September 28 and October 14 (*BZ* **20** (1938) 59).

Position:	20 22 06.26	+ 33 42 18.3	(POSS)
	20 22 06.26	+ 33 42 17.1	(A. A. Wachmann, *BZ* **20** (1938) 59)
	73.119	− 2.091	(G.C.)

Range: 12.5p (11.0) – 20.5p LCT: Bb? t_3: 60^d

Finding chart:	A. A. Wachmann, *Erg AN* **11** (5) (1948) E42.
Light curve:	A. A. Wachmann, *Erg AN* **11** (5) (1948) E42.
Spectroscopy:	R. B. Baldwin, *PAAS* **9** (1938) 33.

Identification: from Wachmann's (1948) finding chart.
Classification: moderately fast nova; in extremely reddened region of the Galaxy.

V407 Cyg — ZAND

(148.1940 Cyg, MHα 289-90)

Discovered by C. Hoffmeister on Sonneberg plates and classified as novalike object. A major outburst occurred in 1936 (*VSS* **1** (1949) 295).

Position:			
	21 00 24.26	+45 34 40.8	(POSS)
	21 00 26	+45 34 36	(C. Hoffmeister, *VSS* **1** (1949) 295)
	86.983	− 0.482	(G.C.)

Range: 13.3p – 17.0p LCT: pec t_3: ?

Finding chart: C. Hoffmeister, *MVS* 327 (1957); L. Meinunger, *MVS* **3** (1966) 111; D. A. Allen (1984).
Light curve: L. Meinunger, *MVS* **3** (1966) 111; F. Gieseking, *AsAp Suppl* **26** (1976) 367.
Spectroscopy: W. Bidelman, *ApJ Suppl* **1** (1954) 207; G. H. Herbig, *ApJ* **131** (1960) 632; V. P. Esipov, B. F. Yudin, *ATs* 1415 (1986), H. W. Duerbeck (unpublished).
Identification: from Allen's finding chart.
Classification: L. Meinunger (*MVS* **3** (1966) 111) finds Mira-type variability with t_{max} = 2429710 + 745 · E, which is, however, not well established (F. Gieseking, *AsAp Suppl* **26** (1976) 376). The brightness rise occurred on 1936 August 29, followed by a steady decline; it is traceable until 1939 August. Duerbeck and Esipov and Yudin find the spectrum of a symbiotic star.

V450 Cyg — NB

(N Cyg 1942, 85.1942 Cyg)

Discovered by F. Zwicky, Palomar Observatory, as an 8^m star, 1942 September 8. Maximum occurred between 1942 May 10 and June 4, at about 7^m (*IAU Circ* 918, *HAC* 631).

Position:			
	20 56 48.15	+35 44 46.2	(POSS)
	20 56 48.16	+35 44 47.2	(2 outburst observations)
	79.128	− 6.458	(G.C.)

Range: 7.8p (7.0) – 16.3p? LCT: Ca t_3: 108^d

Finding chart: M. D. Ashbrook, V. McKibben Nail, *HB* 916 (1942) 20; J. Stein, J. Junkes, *Ric Astr* **1** (1945) 337; K. Himpel, *BZ* **24** (1942) 94.

Light curve: M. D. Ashbrook, V. McKibben Nail, *HB* 612 (1942) 20; P. Ahnert *et al.*, *BZ* **24, 25, 26** (1942-1944); J. Stein, J. Junkes, *Ric Astr* **1** (1945) 337; C. Payne-Gaposchkin (1957) 13.

Spectroscopy: R. F. Sanford, *ApJ* **97** (1943) 130 – phot, rv; D. B. McLaughlin, *AJ* **58** (1953) 220 – rv.

Identification: from Harvard plates MC 32822, taken 1943 January 7/8, and IR 6089, IR 6090, taken 1942 September 1/2. The nova is a blend of three stars on the POSS. The exnova is probably the star whose coordinates are listed above. They are very similar to those determined by S. Cederblad (*Lund Ann* **13** (1954) 29). A second candidate is at RA: 20 56 48.28, Decl. +35 44 48.3.

Classification: well-observed moderately fast nova of DQ Her-type with exceedingly deep minimum.

V465 Cyg **NB**

(N Cyg 1948, VV 67, OV 11)

Discovered by B. S. Whitney, Norman, Oklahoma, 1948 June 2, as a 10^m star. It appears first on a plate of 1948 May 31 (9^m) (*IAU Circ* 1154, *HAC* 902).

Position:	19 50 47.60	+ 36 26 03.3	(POSS)
	19 50 47.64	+ 36 26 03.7	(5 outburst observations)
	71.908	+ 4.764	(G.C.)

Range: 8.0 p (7.3) – 17.0p LCT: Bb t_3: 140^d

Finding chart: V. A. Kolychev, *PZv* **8** (1951) 385; E. A. Dibay, *PZv* **12** (1958) 376; W. J. Miller, A. A. Wachmann, *Ric Astr* **6** (1959) 264; L. Rosino, *Bologna Pubbl* **5** (1952) No. 21 = *Mem SA It* **23** (1952) 109.

Light curve: J. Ashbrook, V. McKibben Nail, *AJ* **55** (1949) 95; A. B. Soloviev, Tadj (= Stalinabad) *Tsirk* **67-68** (1949); M. Beyer, *AN* **280** (1951) 273; W. J. Miller, A. A. Wachmann, *Ric Astr* **6** (1959) 264; D. Ya. Martynov, *PZv* **13** (1960) 142; C. Payne-Gaporschkin (1957) 11.

Spectroscopy: M. Bloch, Ch. Fehrenbach, *CR* **227** (1948) 265 – phot, trac, ident, rv; A. H. Joy, *PASP* **60** (1948) 265 – descr, rv; K. M. Yoss, *PASP* **61** (1949) 87 – descr, rv; M. Bloch, *AAp* **13** (1950) 390 – phot, trac, ident, rv.

Identification: from Harvard plates MC 36065, taken 1948 August 10/11, and MC 36076, taken 1948 August 25/26, and from the precise positions determined during outburst.

Classification: well-observed slow nova. The light curve shows several oscillations with ~ 1^m amplitude, followed by a smooth decline.

V476 Cyg NA

(N Cyg 1920, N Cyg No. 3, 26.1920 Cyg)

Discovered by W. F. Denning, Bristol, 1920 August 20, when the nova was $3^m.5$. Maximum brightness was reached on 1920 August 24 (1.6v); the rise occurred between 1920 August 16 and 19 (*AN* **211** (1920) 371).

Position:	19 57 09.64	+ 53 28 54.5	(13 outburst observations)
	19 57 09.52	+ 53 28 54.5	(2 recent observations)
	87.368	+ 12.417	(G.C.)

Range: **2.0**p – 17.2B LCT: A t_3: 16.5^d

Finding chart: M. Humason (1938); A. Sh. Khatisov (1971).
Light curve: L. Campbell, *HB* 890 (1932) 3; C. Bertaud (1945) 84; G. P. Sacharov, *PZv* **9** (1953) 175; G. Cecchini, L. Gratton (1941) 120, 122; C. Payne-Gaposchkin (1957), 9, 111.
Spectroscopy: R. B. Baldwin, *Michigan Publ* **8** (1940) 61 – phot, ident, rv; W. E. Harper, *Victoria Publ* **1** (1920) 267 – phot, rv; W. S. J. Lockyer, D. L. Edwards, *MN* **81** (1921) 173 – phot, rv; F. J. M. Stratton, *MN* **82** (1921) 44 – wavelengths; J. Storey, *MN* **81** (9120) 141 – descr, phot; C. Bertaud (1945) 84; C. Payne-Gaposchkin (1957) 111.
Nebular shell: W. Baade, *PASP* **56** (1944) 218; E. R. Mustel, A. A. Boyarchuk, *ApSS* **6** (1970) 183; H. W. Duerbeck, *ApSS* **131** (1987) 461.
Identification: from published finding charts.
Classification: fast nova, observed almost to minimum light.

V1330 Cyg NA

(N Cyg 1970)

Discovered by F. M. Stienon, Warner and Swasey Observatory, on an objective prism plate taken 1970 June 8 (*IAU Circ* 2251).

Position:	20 50 46.46	+ 35 48 04.4	(POSS)
	20 50 46.36	+ 35 48 02.1	(2 outburst observations)
	78.376	– 5.488	(G.C.)

Range: 9p (7.5) – 18.1p LCT: Ao? t_3: 18^d

Finding chart: A. Sh. Khatisov, *ATs* 796 (1973).
Light curve: F. Ciatti, L. Rosino, *AsAp Suppl* **16** (1974) 305.
Spectroscopy: F. Ciatti, L. Rosino, *AsAp Suppl* **16** (1974) 305.
Duplicity: visual companion of similar brightness at RA 20 50 46.98, Decl + 35 47 58.4.
Identification: from Khatisov's finding chart and published precise positions.
Classification: fast nova; maximum not covered by observations.

V1500 Cyg NA

(N Cyg 1975)

Discovered by K. Osada, Japan, 1975 August 29.48, as a star of 3^m. Maximum light was reached on 1975 August 31 (*IAU Circ* 2826).

Position: 21 09 52.95 + 47 56 40.95 (POSS blue, extremely faint star)
21 09 52.857 + 47 56 41.05 (14 outburst observations)
89.823 − 0.073 (G.C.)

Range: **2.2**B − 21.5p LCT: A t_3: 3.6^d

Finding chart: *ATs* 889 (1975) 5; W. R. Beardsley, M. W. King, J. L. Russell, J. W. Stein, *PASP* **87** (1975) 943.
Light curve: Z. Alksne, I. Platais, *ATs* 889 (1975) 8 (prediscovery); G. A. Richter, *MVS* **7** (1975) 46 (maximum); P. J. Young, A. G. Corwin, jr., J. Bryon, G. de Vaucouleurs, *ApJ* **209** (1976) 882; K. Ichimura, M. Nakagiri, E. Watanabe, K. Okida, S. Nishimura, Y. Yamashita, *Tokyo Bull 2nd Ser* **241** (1975) 2055; P. Tempesti, *AN* **300** (1979) 51.
Spectroscopy: D. J. Stickland, *ApSS* **92** (1983) 197 – premaximum spectra; A. A. Boyarchuk, T. S. Galkina, R. E. Gershberg, V. I. Krasnobabtsev, T. M. Radkovskaya, N. I. Shakovskaya, *AZh* **54** (1977) 458 = *Sov Astr* **21** (1977) 257 – phot, ident, trac; L. Rosino, P. Tempesti, *AZh* **54** (1977) 517 = *Sov Astr* **21** (1977) 291 – phot, ident, trac; H. W. Duerbeck, B. Wolf, *AsAp Suppl* **29** (1977) 291 – ident, trac, rv; J. B. Hutchings, J. E. Bernard, L. Margentish, M. McCall, *Victoria Publ* **15** (1978) 73 – trac, rv; Ch. Fehrenbach, Y. Andrillat, *CR* **281** (1975) 365 – phot, rv; E. Kontizas, M. Kontizas, M. J. Smyth, *MN* **176** (1976) 79p – spectrophot; G. Ferland, D. L. Lambert, J. H. Woodman, *ApJ* **213** (1977) 137 – coronal lines; G. Ferland, D. L. Lambert, J. H. Woodman, *ApJ Suppl* **60** (1986) 375 – trac, spectrophot; Yu. V. Borisov, S. I. Gerasimenko, *Tadj Byull* **75** (1984) 1, 16 – spectrophotometry, tables.
IR observations: J. S. Gallagher, E. P. Ney, *ApJ* **204** (1976) L35; V. I. Shenavrin, V. I.

	Moroz, A. A. Liberman, *Pisma AZh* **2** (1976) 94; D. Ennis, E. E. Becklin, S. Beckwith, J. Elias, I. Gatley, K. Matthews, G. Neugebauer, S. P. Willner, *ApJ* **214** (1977) 478.
UV observations:	E. B. Jenkins, T. P. Snow, W. L. Upson, S. G. Starrfield. J. S. Gallagher, M. Friedjung, J. L. Linsky, R. Anderson, R. C. Henry, H. W. Moos, *ApJ* **212** (1977) 198; C. C. Wu, D. Kester, *AsAp* **58** (1977) 331.
Radio observations:	V. I. Altunin, *Pisma AZh* **2** (1976) 299; E. R. Seaquist, N. Duric, F. P. Israel, T. A. T. Spoelstra, B. L. Ulich, P. C. Gregory, *AJ* **85** (1980) 283; R. M. Hjellming, C. M. Wade, N. R. Vandenberg, R. T. Newell, *AJ* **84** (1979) 1619.
Polarimetry:	V. Piirola, *AZh* **54** (1977) 612.
Duplicity:	(short period variations): P. Tempesti, *IBVS* 1052, *IBVS* 1057 (1975); M. Marcocci, I. Mazzitelli, R. Messi, G. Natali, R. Rossi, *AsAp* **55** (1977) 171; I. Semeniuk, A. Kruszewksi, A. Schwarzenberg-Czerny, T. Chlebowski, M. Mikolajewski, A. Wosczcyk, *AA* **27** (1977) 301; H. H. Lanning, I. Semeniuk, *AA* **31** (1981) 175; J. B. Hutchings, M. L. McCall, *ApJ* **217** (1977)) 775; J. Patterson, *ApJ* **225** (1978) 954, *ApJ* **231** (1979) 789; A. Kruszewski, I. Semeniuk, H. W. Duerbeck, *AA* **33** (1983) 339; E. P. Pavlenko, *ATs* 1239 (1982) 3.
Nebular shell:	H. J. Becker, H. W. Duerbeck, *PASP* **92** (1980) 792.
Identification:	from published precise positions; prenova fairly well visible on POSS glass copy (blue).
Classification:	very well observed very fast nova; large outburst amplitude.

V1668 Cyg — NA

(N Cyg 1978)

Discovered by W. Morrison, Peterborough, Ontario, and P. L. Collins, Mt. Hopkins Observatory, 1978 September 10 (*IAU Circ* 3263, 3264).

Position:	21 40 38.06	+ 43 48 10.1	(POSS)
	21 40 38.18	+ 43 48 10.2	(3 outburst observations)
	90.838	− 6.760	(G.C.)

Range: **6.7**p – 20.0p LCT: Ba t_3: 23^d

Finding chart:	H. W. Duerbeck, H. Pollok, *IBVS* 1845 (1980).
Light curve:	H. W. Duerbeck, K. Rindermann, W. C. Seitter, *AsAp* **81** (1980) 157; J. S. Gallagher, J. B. Kaler, E. C. Olson, W. I. Hartkopf, D. A. Hunter, *PASP* **92** (1980) 46; A. D. Paolantonio, R. Patriarca, P.

Tempesti, *IBVS* 1913 (1981); A. D. Mallama, D. R. Skillman, *PASP* **91** (1979) 91; W. Blitzstein, D. H. Bradstreet, B. J. Hrivnak, A. B. Hull, R. H. Koch, R. J. Pfeiffer, A. P. Galatola, *PASP* **92** (1980) 338; J. A. Mattei, *JRAS Can* **74** (1980) 185; J. B. Kaler, *PASP* **98** (1986) 243.

Spectroscopy: G. Klare, B. Wolf, J. Krautter, *AsAp* **89** (1980) 282 – trac, rv; S. E. Smith, P. V. Noah, M. J. Cottrell, *PASP* **91** (1979) 775 – phot, trac, rv; Ch. Fehrenbach, Y. Andrillat, *CR Sér B* **228** (1979) 191 – phot, rv.

Polarization: V. Piirola, T. Korhonen, *AsAp* **79** (1979) 254.

UV observations: A. Cassatella, P. Benvenuti, J. Clavel, A. Heck, M. V. Penston, P. L. Selvelli, F. Macchetto, *AsAp* **74** (1979) L18; M. Friedjung, *AsAp* **93** (1981) 320; D. J. Stickland, C. J. Penn, M. J. Seaton, M. A. J. Snijders, P. L. Storey, *MN* **197** (1981) 107.

IR observations: J. P. Phillips, R. Wade, M. J. Selby, D. Sanchez Magro, *MN* **187** (1979) 45p; R. D. Gehrz, J. A. Hackwell, G. L. Grasdalen, E. P. Ney, G. Neugebauer, K. Sellgren, *ApJ* **239** (1980) 570.

Duplicity (short period variations): F. Campolonghi, R. Gilmozzi, R. Messi, G. Natali, J. Wells, *AsAp* **85** (1980) L4; I. B. Voloshina, A. M. Cherepashchuk, *ATs* 1042 (1980) 2; A. Piccioni, A. Guarneri, C. Bartolini, F. Giovanelli, *AA* **34** (1984) 473.

Identification: from Harvard plate MC 40022, taken 1978 September 10.

Classification: fast nova, well observed from the ultraviolet to the infrared. At maximum and during the declining phase, light variations with P = 0.4271 d were observed (H. Ritter 1984).

V1760 Cyg **M**

(N Cyg 1980, Honda's variable)

Discovered by M. Honda, Japan, 1980 November 29, as a 10^{m} star (*IAU Circ* 3546).

Position:	21 40 46.19	+ 31 13 45.3	(POSS)
	21 40 46.18	+ 31 13 45.05	(2 outburst observations)
	82.325	− 16.177	(G.C.)

Range: 11.8p – [15p LCT: M t_3: –

Identification: from published precise positions.

Classification: Mira variable with P = 298^{d} (E. Waagen, *IAU Circ* 3553 (1980))

N Cyg 1986 — NB

Discovered by M. Wakuda, Japan, 1986 August 4.7, as a star of $9^{m}.4$v. It was 13v on 1986 July 28.6 (*IAU Circ* 4242).

Position:	19 52 45.90	+ 35 34 20.7	(POSS, nearest star)
	19 52 45.885	+ 35 34 18.7	(2 outburst observations)
	71.372	+ 3.978	(G.C.)

Range: 8.7v – 19p   LCT: Bb? t_3: > 100^d

Finding chart: R. Chanal, *BAFOEV* **38** (1986) 5.
Light curves: E. Schweitzer, *BAFOEV* **38** (1986) 5; M. Wakuda, M. Huruhata, *IBVS* 2933 (1986).
Spectroscopy: Y. Andrillat, L. Houziaux, *IAU Circ* 4260 (1986) – descr; C. Aikman, H. Kosai, L. Rosino, T. Iijima, *IAU Circ* 4246 (1986) – descr.
IR observations: R. D. Gehrz, *IAU Circ* 4259 (1986).
Identification: from published precise positions.
Classification: slow nova.

Q Cyg — NA

(N Cyg 1876, N Cyg No. 2, BD +42°4182[a]

Discovered by J. Schmidt, Athens, on 1876 November 24, as a 3^m star. Until 1876 November 20, no star]5^m was present. The nova remained at 3^m until November 27 (*AN* **89** (1876) 9).

Position:	21 39 45.38	+ 42 36 45.6	(A. Sh. Khatisov (1971))
	21 39 45.34	+ 42 36 45.6	(3 outburst observations)
	89.928	– 7.552	(G.C.)

Range: **3.0**v – 15.6v   LCT: A t_3: 11^d

Finding chart: M. Humason (1938); A. Sh. Khatisov (1971).
Light curve: J. Schmidt, *AN* **89** (1877) 41; N. Lockyer, *Phil Trans* **182A** (1891) 397; G. Cecchini, L. Gratton (1941) 36; C. Payne-Gaposchkin (1957) 9; S. Yu. Shugarov, *PZv* **21** (1983) 807 – minimum light.
Spectroscopy: J. N. Lockyer, *Phil Trans* **182A** (1891) 397; H. Vogel, K. Lohse, *MB PrAW* (1877) 241, 826, *MB PrAW* (1878) 302; A. Cornu, *CR* **83** (1877) 1098, 1172; A. Secchi, *CR* **84** (1877) 296; P. Backhouse, *MN* **39** (1879) 34.
Identification: from published finding charts.
Classification: very fast nova.

HR Del

NB

(N Del 1967)

Discovered by G. E. D. Alcock, Peterborough, England, 1967 July 8.9, as a 5^{m} star (*IAU Circ* 2022).

Position:	20 40 04.11	+ 18 58 52.2	(POSS)
	20 40 04.19	+ 18 58 51.0	(A. Sh. Khatisov (1971))
	63.431	− 13.972	(G.C.)

Range: **3.5**v – 12.0v LCT: D t_3: 230^{d}

Finding chart: G. B. Stephenson, *PASP* **79** (1967) 586; A. Sh. Khatisov (1971).

Light curve: A. Terzan, *AsAp* **5** (1970) 167; J. E. Isles, *JBAA* **85** (1974) 54; H. Drechsel, J. Rahe, H. W. Duerbeck, L. Kohoutek, W. C. Seitter, *AsAp Suppl* **30** (1977) 323; E. L. Robinson, *AJ* **80** (1975) 575 – pre-outburst.

Spectroscopy: Ch. Fehrenbach, Y. Andrillat, M. Bloch, *CR Sér B* **263** (1967) 583 – phot, trac, ident, rv; *CR* **265** (1967) 1149 – phot, descr, rv; *CR* **267** (1968) 1177 – phot. ident; *CR* **269** (1969) 546 – phot, trac, ident; Y. Andrillat, Ch. Fehrenbach, *ApSS* **78** (1981) 149 – phot; Ch. Fehrenbach, M. Petit, *AsAp* **1** (1969) 403 – phot, rv; J. B. Hutchings, *Victoria Publ* **13** (1969) 397 – trac, rv; W. C. Seitter, *IAU Coll. 15 = Bamb Ver* **9**, No. 100 (1971) 268 – trac, rv; H. Drechsel, J. Rahe, H. W. Duerbeck, L. Kohoutek, W. C. Seitter, *AsAp Suppl* **30** (1977) 323 – phot, trac, spectrophotometry; W. Götz, *MVS* **58** (1970) L1 – pre-outburst spectrum.

Radio observations: R. M. Hjellming, C. M. Wade, *ApJ* **182** (1970) L1; R. M. Hjellming, C. M. Wade, N. R. Vandenberg, R. T. Newell, *AJ* **84** (1979) 1619.

Polarization: B. Zellner, M. D. Morrison, *AJ* **76** (1971) 645; J. Arsenievic, A. Kubicela, *IBVS* 495 (1970); E. T. Belonkon', O. S. Shulov, *AO LGU Trudy* **30** (Ser. Mat. Nauk 50, 1974) 103.

Nebular Shell: D. Soderblom, *PASP* **88** (1976) 517; L. Kohoutek, *MN* **196** (1981) 87p; J. Solf, *ApJ* **273** (1983) 647.

Duplicity: spectroscopic binary, P = 0.2141674 (A. Bruch, *PASP* **94** (1982) 912) supersedes all previous determinations (H. Ritter (1984)).

UV observations: J. Krautter, G. Klare, B. Wolf, H. W. Duerbeck, J. Rahe, N. Vogt, W. Wargau, *AsAp* **102** (1981) 337; J. B. Hutchings, *PASP* **92** (1980) 458.

Identification: from published finding charts.

Classification: well-observed slow nova with extended pre-maximum halt (W. C. Seitter, *IAU Coll. 4*, Budapest 1969, p. 277).

SY Gem N??

(N Gem 1856b, 1.1908 Gem, BD +31°1380)

Observed for the BD, 1857 February 16 ($9^{m}.3$) and 1858 January 22 ($9^{m}.2$); seen by S. Enebo, 1906 December 24 ($9^{m}.5$), and by J. F. Schroeter, 1904 April 18 ($9^{m}.5$), otherwise always [$10^{m}.0$ or $12^{m}.0$ (*AN* **177** (1908) 74).

Position:			
(1)	06 37 23.49	+31 14 37.2	(POSS; A. Sh. Khatisov's (1971) identification)
(2)	06 37 24.02	+31 12 21.9	(POSS; variable found on Potsdam plate, 1896 February 8)
	06 37 23	+31 14 16	(BD position)
	183.637	+11.552	(G.C.)

Range: 9.2v – [13v LCT: ? t_3: ?

Finding chart: A. Sh. Khatisov (1971).
Identification: Khatisov's star is in the BD 'error box'; it is not particularly blue.
Classification: a doubtful object, because confusion with neighbouring stars cannot be excluded. P. P. Parenago (*PZv* **4** (1933) 228) always found the object [14^{m} and suspects that Enebo and Schroeter observed the wrong star. He classifies it as a nova or dwarf nova. GCVS4 classification is N:.

VZ Gem N??

(N Gem 1856a, BD +31°1736)

Observed for the BD, 1856 March 31 ($8^{m}.7$) and 1856 April 2 ($9^{m}.2$). Reported missing by R. and W. Luther (*AN* **180** (1909) 247).

Position:			
(1)	08 04 42.65	+30 58 45.9	(POSS, 74″ from BD position)
(2)	08 04 34.98	+31 00 23.7	(POSS, 65″ distant)
(3)	08 04 34.18	+31 00 26.8	(POSS, 75″ distant)
(4)	08 04 45.68	+31 00 30.4	(POSS, 105″ distant)
(5)	08 04 36.14	+30 58 04.5	(POSS, 98″ distant)
	08 04 39	+30 59 38	(BD position)
	190.899	+28.924	(G.C.; BD position)

Range: 8.7v – [21? LCT: ? t_3: ?

Identification: ambiguous.
Classification: P. P. Parenago (*PZv* **4** (1933) 229) suspects nova. GCVS4 classification is N:.

CI Gem **UG?**

(101.1943 Gem, S 3428)

Discovered by C. Hoffmeister on Sonneberg plates a a dwarf nova or novalike star. Maximum light occurred 1940 January 3 (*MVS* 30 (1943), *AN* **274** (1943) 37).

Position:	06 27 04.71	+ 22 21 02.1	(POSS)
	06 26 56	+ 22 20 40	(C. Hoffmeister, *AN* **274** (1943) 36)
	190.664	+ 5.542	(G.C.)

Range: **14.7**p – 18.5p LCT: ? t_3: ?

Finding chart: C. Hoffmeister, *MVS* 278 (1957).

Light curve: P. Ahnert, C. Hoffmeister, E. Rohlfs, A. van der Voorde, *VSS* **1** (1947) 107.

Identification: from Hoffmeister's finding chart; blue star, fairly certain identification.

Classification: dwarf nova? No spectroscopic observations are available; possibly small outburst amplitude. Only one maximum was observed. GCVS4 classification: N:; M. Petit (*JO* **43** (1960) 17, *JO* **44** (1961) 6) includes it in his catalogue of dwarf novae.

DM Gem **NA**

(N Gem 1903, N Gem No. 1, 12.1903 Gem, HD 48328)

Discovered by H. H. Turner, Greenwich Observatory, on a Carte du Ciel plate taken 1903 March 16. Harvard plates show that the nova was [9^m on 1903 March 2 and $5^m.1$ on March 6 (*AN* **161** (1903) 307; *Obs* **26** (1903) 226).

Position:	06 41 00.51	+ 29 59 47.1	(POSS)
	06 41 00.69	+ 29 59 47.3	(9 outburst observations)
	185.127	+ 11.728	(G.C.)

Range: **4.8**v – 16.7p LCT: A or Ba t_3: 22^d

Finding chart: M. Humason (1938); A. Sh. Khatisov (1971) – Humason's chart and Khatisov's position refer to the W companion of the nova.

Light curve: H. Leavitt, *HA* **84** (1920) 121; G. Cecchini, L. Gratton (1941) 74, 75; C. Payne-Gaposchkin (1957) 9.

Spectroscopy: H. D. Curtis, *ApJ* **19** (1904) 83 – vis descr; G. E. Hale, *ApJ* **17** (1903) 303 – vis descr; C. D. Perrine, *ApJ* **18** (1903) 297, *ApJ* **19** (1904) 80

– descr; H. M. Reese, R. H. Curtiss, *ApJ* **18** (1903) 299 – trac, ident; A. J. Cannon, *HA* **76** (1916) 19 – descr.

Identification: from published positions.

Classification: maximum poorly covered; otherwise well-observed fast nova.

DN Gem NA

(N Gem 1912, N Gem No. 2, 18.1912 Gem, HD 50480)

Discovered by S. Enebo, Dombaas, Norway, 1912 March 12, as a star of $4^{m}.2$. It was [11^{m} on March 10 and 5^{m} on March 11 (*AN* **191** (1912) 31).

Position:	06 51 39.72	+ 32 12 18.8	(2 recent observations)
	06 51 39.79	+ 32 12 19.4	(22 outburst observations)
	184.018	+ 14.714	(G.C.)

Range: **3.5**p – 15.8p LCT: Bb t_3: 37^{d}

Finding chart: M. Humason (1938); A. Sh. Khatisov (1971); G. Williams (1983).

Light curve: J. Fischer-Petersen, *AN* **192** (1912) 429; L. Campbell, *HA* **76** (1915) 192; H. Leavitt, *HA* **84** (1920) 121; G. Cecchini, L. Gratton (1941) 94, 95, 96; C. Bertaud (1945) 45; C. Payne-Gaposchkin (1957) 11, 114; O. D. Dokuchaeva, *PZv* **12** (1958) 358.

Spectroscopy: H. Giebeler, *AN* **191** (1912) 393 – trac, ident; A. J. Cannon, *HA* **76** (1912) 393 – phot, descr; F. J. M. Stratton, *ASPO Camb* **4** (1920) 1 – phot, descr, rv; W. H. Wright, *Lick Publ* **14** (1926) 27 – ident, descr, rv, phot; C. Bertaud (1945) 45 – rv; D. B. McLaughlin, *ApJ* **117** (1953) 279, *ApJ* **118** (1953) 27, *AAp* **27** (1964) – rv.

Identification: from published finding charts.

Classification: well-observed fast nova.

N Gem 1892 N??

(NSV 03313)

Observed visually by E. E. Barnard with the 36″ refractor of the Lick Observatory, 1892 August 13, as a 7^{m} star (*AN* **172** (1906) 25. M. Wolf did not find a star]14^{m} on Heidelberg plates taken 1892 December 26; no Harvard plates are available for the year 1892.

Position:	(1) 06 55 49.99	+ 17 06 17.8	(POSS)
	(2) 06 55 50.35	+ 17 06 18.0	(POSS)
	(3) 06 55 50.30	+ 17 06 22.6	(POSS)

(4) 06 55 50.89	+ 17 06 18.9	(POSS)
(5) 06 55 51.97	+ 17 06 25.0	(POSS)
06 55 51	+ 17 06 19	(E. E. Barnard, *AN* **172** (1906) 25)
198.448	+ 9.248	(G.C.)

Range: 7.0v – [14p LCT: ? t_3: ?

Identification: Unclear. Some stars near Barnard's position are listed under Nos. 1–5.
Classification: seen only once; doubtful object.

DQ Her NA

(N Her 1934, 452.1934 Her)

Discovered by Prentice, England, 1934 December 12 (*IAU Circ* 152, *HAC* 318, *BZ* **16**, 77, *AN* **254** (1934) 81).

Position:	18 06 05.28	+ 45 51 02.2	(3 recent observations)
	18 06 05.38	+ 45 51 02.2	(17 outburst observations)
	73.153	+ 26.444	(G.C.)

Range: **1.3**v – 14.5v (var) LCT: Ca t_3: 94^d

Finding chart: A. Sh. Khatisov (1971); G. Williams (1983)
Light curve: A. Beer, *MN* **95** (1935) 538; H. Krumpholz, *Wien Mitt* **1**, 4 (1935) 214; L. Campbell, *HB* 898 (1935) 20; H. Grouiller, *Lyon Publ* **1**, 16 (1936); B. W. Kukarkin, H. K. Gitz, *AZh* **14** (1937) 220; C. Bertaud (1945) 109; G. Cecchini, L. Gratton (1941) 149; C. Payne-Gaposchkin (1957) 13, 14, 126; E. L. Robinson, *AJ* **80** (1975) 515 – pre-outburst.
Spectroscopy: F. J. M. Stratton, *ASPO Camb* **4** (1936) 133 – phot, ident, rv; D. B. McLaughlin, *Michigan Publ* **6** (1937) 107 – phot, ident, rv; F. J. M. Stratton, W. H. Manning, Atlas of Spectra of Nova Herculis, Cambridge (1939) – phot; C. Bertaud (1945) – rv;
C. Sneden, D. L. Lambert, *MN* **170** (1975) 533 – CN isotopes; E. R. Mustel, L. I. Baranova, *AZh* **42** (1965) 42 = *Sov Astr* **9** (1965) 31 – abundances, L. I. Antipova, *AZh* **46** (1969) 366, *AZh* **48** (1971) 288 = *Sov Astr* **13** (1969) 288, *Sov. Astr* **15** (1971) 225 – CN bands, H lines;
R. P. Kraft, *ApJ* **139** (1964) 457 – minimum spectrum, phot.
UV observations: L. Hartmann, J. Raymond, in 'The Universe at Ultraviolet Wavelengths', *NASA-CP 2171* (1981), ed. R. D. Chapman, p. 495.

Nebular shell: E. R. Mustel, A. A. Boyarchuk, *ApSS* **6** (1970) 183 – photography; R. E. Williams, N. J. Woolf, E. K. Hege, R. L. Moore, D. A. Kopriva, *ApJ* **224** (1978) 171 – optical spectroscopy; G. J. Ferland, R. E. Williams, D. L. Lambert, G. A. Shields, M. Slovak, P. M. Gondhalekar, J. W. Truran, *ApJ* **281** (1984) 194 – uv spectroscopy; H. Itoh, *PASJ* **33** (1981) 743 – model; G. J. Ferland, J. W. Truran, *ApJ* **244** (1981) 1022 – model.

Duplicity: spectroscopic and eclipsing binary with P = 0.193621 d: M. F. Walker, *ApJ* **123** (1956) 68, *ApJ* **127** (1958) 319; J. B. Hutchings, A. P. Cowley, D. Crampton, *ApJ* **232** (1978) 500; J. Smak, *AA* **30** (1980) 267; 71^{m} pulsation: S. Balachandran, E. L. Robinson, S. O. Kepler, *PASP* **95** (1983) 653; visual binary; J. Patterson, *PASP* **91** (1979) 487.

Identification: from published finding charts.

Classification: well-observed moderately fast nova with pronounced dust formation; unusual occurrence of molecular lines in the premaximum spectrum.

V360 Her N?

(N Her 1892, PR 1230, KZP 101642)

Discovered by J. Baillaud and P. de Grandchamp, Observatoire de Paris, on a plate taken 1892 July 8, as a star of $6^{m}.3$ (*JO* **10** (1927) 125).

Position:			
	17 14 33.88	+ 24 30 04.4	(POSS)
	17 14 34.19	+ 24 30 00.2	(J. Baillaud, P. de Grandchamp, *JO* **10** (1927) 125)
	46.520	+ 30.957	(G.C.)

Range: 6.3p – ? LCT: ? t_3: ?

Identification: best candidate on POSS is 6″ from published position. This object is star No. 244 on Paris photograph $+24° 17^{h}12^{m}$, and in the Catalogue photographique du ciel, Observatoire de Paris, Coordonnées Rectilingues, Tome I, Zones + 23° à + 25°, p. A157 (Paris 1902). J. Ashbrook examined Harvard plates: 1892 June 13, [11^{m}, 1893 June 29, [11^{m} (*AJ* **58** (1953) 176).

Classification: dubious object. A study of the candidate star would be helpful.

V446 Her NA

(N Her 1960)

Discovered by O. Hassel, Oslo, 1960 March 7, as a 5^{m} star (*IAU Circ* 1714).

Position:	18 55 03.03	+ 13 10 26.6	(POSS)
	18 55 03.08	+ 13 10 26.4	(5 outburst observations)
	45.409	+ 4.707	(G.C.)

Range: **3.0**p – 15.0 ... 18.0 LCT: A t_3: 16^d

Finding chart: T. Cragg, *PASP* **72** (1960) 472.

Light curve: C. Bertaud, *JO* **45** (1962) 321; K. Gyldenkerne, V. Meydahl, R. M. West, *København Publ 201* (1961); E. L. Robinson, *AJ* **80** (1975) 515 – pre-outburst.

Spectroscopy: Ch. Fehrenbach, *CR* **250** (1960) 2132 – descr, ident, rv; S. Weniger, *CR* **250** (1960) 4105 – phot, trac, ident, descr; J. Dufay, M. Bloch, D. Chalonge, *CR* **251** (1960) 1969 – phot, descr; J. Dufay, M. Bloch, Y. Andrillat, *CR* **251** (1960) 2289 – trac, descr; E. R. Mustel, I. M. Kopylov, L. S. Galkina, R. N. Kumaigorodskaja, T. M. Bartasch, *Izv KrAO* **26** (1961) 181 – phot, trac, ident, line intensities; V. V. Prokofyeva, T. S. Belyakina, *Izv KrAO* **29** (1963) 278 – spectrophotometry; Y. Andrillat, *AAp* **27** (1964) 486 – IR spectrophotometry, phot, trac, ident; A. B. Meinel, *ApJ* **137** (1963) 834 – spectrophotometry; J. Dufay, M. Bloch, D. Chalonge, *AAp* **27** (1964) 539 – phot, trac, ident, rv; M. V. Saveljeva, *AZh* **44** (1967) 716 = *Sov Astr* **11** (1967) 576 – spectrophotometry, trac; G. J. Ferland, *ApJ* **231** (1979) 781 – He abundance.

Polarization: K. A. Grigorian, R. A. Vardanian, *Byurakan Soob* **29** (1961) 39.

Identification: from Cragg's finding chart.

Classification: well-observed fast nova.

V533 Her **NA**

(N Her 1963)

Discovered by E. Dahlgren, Vikmanshyattan, Sweden, 1963 February 6, as a star of $3\overset{m}{.}9$ (*IAU Circ* 1817).

Position:	18 12 46.38	+ 41 50 22.1	(POSS)
	18 12 46.52	+ 41 50 22.8	(7 outburst observations)
	69.188	+ 24.274	(G.C.)

Range: **3.0**p – 15.0p LCT: Ba t_3: 44^d

Finding chart: C. B. Stephenson, R. B. Herr, *PASP* **75** (1963) 253; A. Sh. Khatisov (1971); G. Williams (1983).

Light curve: L. H. Solomon, *SAO SpR* 244 (1967) 1; I. Almár, E. Illés-Almár, *Bud Mitt* **60** (1966); E. L. Robinson, *AJ* **80** (1975) 515 – pre-outburst.

Spectroscopy: M. Friedjung, M. G. Smith, *MN* **132** (1966) 239 – spectrophotometry; T. M. Bartash, A. A. Boyarchuk, *Izv KrAO* **33** (1965) 173 – spectrophotometry; W. C. Seitter, *Bonn Veröff* **67** (1963) – phot, trac, ident, rv; V. T. Doroshenko, *AZh* **45** (1968) 121 = *Sov Astr* **12** (1968) 95 – spectrophotometry; D. B. McLaughlin, *AAp* **27** (1964) 486 – rv; Y. Andrillat, *AAp* **27** (1964) 475, 486 – IR spectroscopy, phot, trac; G. Chincarini, L. Rosino, *AAp* **27** (1964) 469 – phot, ident; M. Bloch, D. Chalonge, *AAp* **27** (1964) 274 – phot, trac, ident;
G. Williams (1983) – minimum spectrum, trac.

Duplicity: light variations with P = 0.28: d and 63^s (transient) (H. Ritter (1984); J. Patterson, *ApJ* **233** (1979) L13; E. L. Robinson, R. E. Nather, *ApJ* **273** (1983) 255).

Identification: from published finding charts.

Classification: well observed moderately fast nova.

V592 Her UG? or XND?

(N Her 1968, S 10376)

Discovered by G. A. Richter on Sonneberg plates; maximum light occurred 1968 June 30 (*IBVS* 293 (1968)).

Position:			
(1)	16 28 46.76	+21 23 22.5	(CA CCD, star 22^m)
(2)	16 28 46.92	+21 22 56.2	(CA CCD, star 21^m)
	38.819	+39.990	(G.C.)

Range: **12.3**p – 21.5:p LCT: C? t_3: 27^d

Finding chart: G. A. Richter, *IBVS* 293 (1968).

Light curve: G. A. Richter, *IBVS* 293 (1968).

Identification: tentative identification on a CCD frame taken with the Calar Alto 2.2m telescope, using Richter's finding chart. Star (1) is nearer to the indicated position than star (2).

Classification: the object is blue at maximum and is probably no classical nova. The light curve resembles that of the X-ray nova V616 Mon. The debatable GCVS4 classification is NA.

V632 Her UG

(N Her 1967)

Discovered by J. Dorschner, Ch. Friedemann and W. Pfau, Sternwarte Jena, on a plate taken 1967 June 29 (16^m). The object was not visible on May 11 (*ATs* 430 (1967) 1).

Position: 18 17 37.89 +24 31 55.1 (POSS)
18 17 37.9 +24 31 55 (W. Pfau, private communication)
52.067 +17.534 (G.C.)

Range: 15.4p – 21p LCT: DN? t_3: 14^d

Finding chart: W. Pfau, private communication.
Light curve: J. Dorschner, Ch. Friedemann, *AN* **291** (1968) 7.
Identification: from Pfau's position.
Classification: blue at maximum; the light curve form and amplitude suggest dwarf nova; GCVS4 classification: UG.

CP Lac

NA

(N Lac 1936, N Cep 1936, 605.1936 Lac, 605.1936 Cep)

Discovered by K. Gomi, Tokyo, 1936 June 18, as a 4^m star (*IAU Circ* 594).

Position: 22 13 50.45 +55 22 02.9 (2 recent observations)
22 13 50.55 +55 22 02.95 (14 outburst observations)
102.141 − 0.837 (G.C.)

Range: **2.1**v – 16.6p LCT: A t_3: 10^d

Finding chart: A. Sh. Khatisov (1971).
Light curve: C. Bertaud (1945) 146; P. P. Parenago, *PZv* **7** (1949) 109; I. D. Howarth, *JBAA* **88** (1978) 608; G. Cecchini, L. Gratton (1941) 158; C. Payne-Gaposchkin (1957) 9, 130; E. L. Robinson, *AJ* **80** (1975) 515 – pre-outburst.
Spectroscopy: W. S. Adams, R. F. Sanford, O. C. Wilson, *PASP* **48** (1948) 235 – rv; W. E. Harper, J. A. Pearce, C. S. Beals, R. M. Petrie, A. McKellar, *Victoria Publ* **6** (1937) 317 – phot, trac, ident, rv; D. B. McLaughlin, *ApJ* **118** (1953) 27 – coronal line; C. Bertaud (1945) 154 – rv; C. Payne-Gaposchkin (1957) 130 – rv, descr; G. J. Ferland, *ApJ* **231** (1979) 781 – He-abundance.
X-ray observations: R. H. Becker, F. E. Marshall, *ApJ* **244** (1981) L93.
Identification: from Khatisov's finding chart.
Classification: well observed very fast nova.

DI Lac

NA

(N Lac 1910, 137.1910 Lac, HD 214239)

Discovered by T. E. Espin, Walsingham Observatory, England, 1910 December 30, as a star of $8^m.0$. Photographic plates date the outburst between 1910 November 17 and 23, when the nova reached 5^m (*AN* **186** (1911) 523).

Position:	22 33 46.51	+ 52 27 26.1	(POSS)
	22 33 46.575	+ 52 27 26.1	(11 outburst observations)
	103.108	− 4.855	(G.C.)

Range: **4.6**v – 14.9p LCT: Bb t_3: 43^d

Finding chart: M. Humason (1938); A. Sh. Khatisov (1971); G. Williams (1983).

Light curve: N. F. Kalashnikov, B. A. Vorontsov-Velyaminov, *AZh* **16** (1939) 29; H. S. Leavitt, *HA* **84** (1920) 121; G. Cecchini, L. Gratton (1941) 91; C. Payne-Gaposchkin (1957) 11; E. L. Robinson, *AJ* **80** (1975) 515 – pre-outburst.

Spectroscopy: W. H. Wright, *Lick Bull* **6**, 194 (1911) 95 – phot, ident; W. S. Adams, F. G. Pease, *PASP* **27** (1915) 237 – descr; A. J. Cannon, *HA* **76** (1916) 19 – descr; J. Genard, *MN* **92** (1931) 396 – trac; R. P. Kraft, *ApJ* **139** (1964) 457 – minimum spectrum, phot; G. Williams (1983) – minimum spectrum, trac.

IR observations: M. R. Sherrington, R. F. Jameson, *MN* **205** (1983) 265 – photometry at minimum light.

Duplicity: spectroscopic binary with P = 0.543773 d(R. F. Webbink, see H. Ritter (1984)).

Identification: from published finding charts.

Classification: moderately fast nova.

DK Lac NA

(N Lac 1950)

Discovered by C. Bertaud, Observatoire de Paris, on a photographic plate taken 1950 January 23, as a star of $6^m.1$. It was [$13^m.5$ on January 18, $6^m.6$ on January 20 (*IAU Circ* 1254).

Position:	22 47 40.455	+ 53 01 24.45	(A. Sh. Khatisov (1971))
	22 47 40.49	+ 53 01 24.45	(4 outburst observations)
	105.237	− 5.352	(G.C.)

Range: 5.9 p (5.0p) – 15.5p LCT: Ao? Ba? t_3: 32^d

Finding chart: G. Larsson-Leander, *Stockh Ann* **18**, 3 (1954); A. Sh. Khatisov (1971).

Light curve: G. Larsson-Leander, *Stockh Ann* **18**, 3 (1954); M. Beyer, *AN* **280** (1951) 273; J. Ribbe, *PASP* **63** (1951) 39; C. Bertaud, F. Baldet, *JO* **35** (1952) 108; Z. Bochnicek, *BAC* **2** (1950) 88; M. Schmidt, *BAN* **11** (1950) 244; Ch. Fehrenbach, A. Duflot, *JO* **33** (1950) 54; C. Payne-Gaposchkin (1957) 11, 132, 133.

Spectroscopy: G. Larsson-Leander, *Stockh Ann* **17**, 8 (1953) – trac, ident, rv; G. Larsson-Leander, *Stockh Ann* **18**, 4 (1954) – trac, ident, rv, line intensities; P. Wellmann, *ZsAp* **29** (1951) 112 – ident, rv, line intensities; M. Barbière, Y. Ribelaygue, G. Courtès, Ch. Fehrenbach, *CR* **230** (1950) 1836 – phot, ident, rv; D. B. McLaughlin, *ApJ* **118** (1953) 27 – coronal line; D. B. McLaughlin, *AAp* **27** (1964) 450 – rv.

Identification: from Heidelberg plates B 7185/7186, taken 1950 February 17.

Classification: well observed fast nova; light curve shows many unusually strong oscillations.

N Lac 1986 N?

Discovered by M. Honda, Japan, 1986 November 22, as an 8^m star. No object $]16^m$ was found on November 25 (*IAU Circ* 4275, 4276).

Position:	22 22 07.6	+ 48 12 41	(H. Kosai, *IAU Circ* 4276 (1986))
	99.298	− 7.504	(G.C.)

Range: 8v – [21? LCT: ? t_3: ?

Identification: from precise position (H. Kosai, *IAU Circ* 4276 (1986); empty field on POSS. Field map in Appendix.

Classification: existence of object not established. The discovery image is seen on two simultaneous exposures.

U Leo N??

(N Leo 1855, BD + 14°2239)

Reported faint or missing by C. H. F. Peters (*AN* **87** (1876) 271). According to H. Kreutz (*AN* **100** (1881) 317), the first of the two BD observations, 1854 January 22, is doutbful, the second one, 1855 January 18, is certain. The Marktree catalogue, which contains observations of this zone made between 1855 March 15 and 1856 March 13, does not list the star.

Position:	10 21 22.7	+ 14 15 24	(BD)
	10 21 23.195	+ 14 15 37.69	(A. Sh. Khatisov (1971))
	226.341	+ 53.263	(G.C.)

Range: 10.5v – [15v LCT: ? t_3: ?

Identification: according to Khatisov's finding chart.

Classification: the reality of the object is not established, the GCVS4 classification is N:.

RZ Leo **UGWZ**

(30.1919 Leo)

Discovered by M. Wolf on Heidelberg plates taken 1918 March 13, as a star of $10^{m}.5$. A second outburst was observed in 1985 (*AN* **209** (1919) 85).

Position:	11 34 48.49	+ 02 05 34.6	(POSS)
	11 34 48.89	+ 02 05 32.0	(M. Mündler, *AN* **209** (1919) 65)
	264.774	+ 59.087	(G.C.)

Range: 10.5p – 17.5p LCT: DN t_3: –

Finding chart: G. H. Herbig, *PASP* **70** (1958) 605; A. Sh. Khatisov's chart is incorrect.
Light curve: S. Cristiani, H. W. Duerbeck, W. C. Seitter (in preparation).
Spectroscopy: S. Cristiani, H. W. Duerbeck, W. C. Seitter, *IAU Circ* 4027 (1985).
Identification: from Heidelberg plates B 1073/1074, taken 1918 March 13.
Classification: outburst spectra indicate dwarf nova. M. Petit (*JO* **42** (1960) 17, *JO* **44** (1961) 275) includes RZ Leo in his catalogue of dwarf novae. The GCVS4 classification, NR, is incorrect.

N Leo 1612 **N??**

(NSV 04550, Zi 767)

Ch. Scheiner observed a star close to Jupiter between 1612 March 30 and April 12 (Apelles, de maculis solaribus et stellis circa Jovem errantibus accuratior descriptio; Aug. Vind. 1612). A. G. Pingré (Annales Célestes, ms. 1786, ed. Bigourdan) assumes that the object is a nova. A. Winnecke (*VJS* **13** (1978) 283) and J. G. Hagen (*ApJ* **17** (1903) 283) assume that it is BD + 15°2083.

Position:		09 34 15	+ 15 28 36	(Ch. Scheiner/A. Winnecke)
	(1)	09 34 14.85	+ 15 28 11.7	(nearest star on POSS)
	(2)	09 34 33.05	+ 15 28 39.25	(BD + 15°2083)
		217.423	+ 43.441	(G.C.)

Range: 4v – ? LCT: ? t_3: ?

Identification: uncertain. The POSS red plate shows 4 additional bright stars in the vicinity which are invisible on the blue plate. They are obviously plate defects. The finding chart in the Atlas covers 8′.6 × 8′.6.
Classification: The CSV states that the object is probably not BD + 15°2083, but a nova-like variable. The reality of the object is not established.

SS LMi UG: or N:

(N LMi 1980)

Discovered by A. Alksnis and L. Zacs on plates taken with the Riga Schmidt telescope. The object was visible in 1980 April (*IBVS* 1972 (1981)).

Position:	10 31 16.3	+ 31 23 29	(POSS, empty field)
	10 31 19	+ 31 24	(A. Alksnis, L. Zacs, *IBVS* 1972 (1981))
	196.997	+ 59.859	(G.C.)

Range: 15.9p – [21p LCT: ? t_3: ?

Finding chart: A. Alksnis, L. Zacs, *IBVS* 1972 (1981).
Light curve: A. Alksnis, L. Zacs, *IBVS* 1972 (1981).
Identification: from Alksnis and Zacs' finding chart. No POSS counterpart.
Classification: no spectroscopic information is available. SS LMi is an extragalactic nova or an unusual dwarf nova of large amplitude.

GW Lib UG: or N:

(N Lib 1983)

Discovered by L. E. González, Santiago, Chile, 1983 August 10, as a 9^m star (*IAU Circ* 3854).

Position:	15 16 58.00	– 24 49 35.4	(GPO plate, May 1986)
	15 16 58.01	– 24 49 36.4	(SRC)
	15 16 58.03	– 24 49 35.7	(L. E. González, *IAU Circ* 3854 (1983))
	340.707	+ 26.766	(G.C.)

Range: 9p – 18.5p LCT: ? t_3: ?

Spectroscopy: H. W. Duerbeck, W. C. Seitter, *ApSS* **131** (1987) 467 – trac, descr.
Identification: from González's accurate position; blue star.
Classification: no spectroscopic observation at maximum is available. The spectroscopic appearance around minimum (16.6v) resembles that of a dwarf nova at minimum: broad H absorption, narrow H emission, Fe II 5169, 5018, 4924 emission. The amplitude suggests nova.

HR Lyr NA

(N Lyr 1919, 1.1920 Lyr, HV 3251, HD 175268)

Discovered by J. C. Mackie on Harvard plates. Maximum light occurred 1919 December 6; the nova was [$16^m.5$ on December 4 (*HB* 705 (1920)).

Position:	18 51 27.64	+ 29 09 50.0	(POSS)
	18 51 27.66	+ 29 09 50.4	(2 recent observations)
	18 51 27.98	+ 29 09 51.1	(E. Hartwig, *VJS* **55** (1920) 171)
	59.584	+ 12.470	(G.C.)

Range: 6.5p (6.5:) – 15.8p LCT: A? t_3: 74^d

Finding chart: M. Humason (1938); A. Sh. Khatisov (1971); G. Williams (1983).
Light curve: H. Grouiller, *JO* **4** (1921) 44; A. A. Nijland, *BAN* **2** (1925) 231; S. I. Bailey, *HB* 705 (1920); G. Cecchini, L. Gratton (1941) 118.
Spectroscopy: W. S. Adams, A. H. Joy, *PASP* **32** (1920) 154 – descr, rv; W. H. Wright, *PASP* **32** (1920) 167 – descr; A. B. Wyse, *Lick Publ* **14** (1940) 229.
R. P. Kraft, *ApJ* **139** (1964) 457 – minimum spectrum, phot; G. Williams (1983) – minimum spectrum, trac.
Identification: from published finding charts.
Classification: moderately fast nova; maximum poorly covered by observations.

BT Mon NA

(N Mon 1939, N Mon No. 2, 67.1939 Mon)

Discovered by A. A. Wachmann on plates taken with the astrograph of the Hamburger Sternwarte 1939 December 17. On 1939 September 24, it was already 7^m.6 (*BZ* **22** (1940) 9, *HAC* 517 (1939)).

Position:	06 41 15.81	– 01 58 08.85	(POSS)
	06 41 15.78	– 01 58 09.12	(2 recent observations)
	06 41 15.91	– 01 58 08.58	(3 outburst observations)
	213.859	– 2.623	(G.C.)

Range: 8.5 p (5:) –15.5 (var) LCT: ? t_3: ?

Finding chart: E. Bertiau, *Leiden Ann* **20** (1940) 12; A. Sh. Khatisov (1971); G. Williams (1983).
Light curve: F. Whipple, *HB* 912 (1940) 12; E. Bertiau, *Leiden Ann* **20** (1954) 358; A. A. Wachmann, *Bergd Abh* **7**, 8 (1968) 387; E. L. Robinson, *AJ* **80** (19756) 515 – pre-outburst.
Spectroscopy: F. Whipple, *HB* 912 (1940) 12 – descr; R. F. Sanford, *PASP* **52** (1940) 35 – phot, trac, ident; P. Swings, O. Struve, *PASP* **53** (1941) 37 – descr.
G. Williams (1983) – minimum spectrum, trac.
Nebular shell: T. R. Marsh, R. A. Wade, J. B. Oke, *MN* **250** (1983) 33p.

Duplicity: eclipsing binary with P = 0.3338141 d; E. L. Robinson, R. E. Nather, S. O. Kepler, *ApJ* **254** (1982) 646 – photometry; W. C. Seitter, *ApSS* **99** (1984) 95 – spectroscopy; B. E. Schaefer, J. Patterson, *ApJ* **268** (1983) 710 – period study.

Identification: from published finding charts.

Classification: probably fast nova. At minimum, deep eclipses, complicated rv curve.

GI Mon **NA**

(N Mon 1918, N Mon No. 1, 2.1918 Mon, HD 58756)

Discovered by M. Wolf on a Heidelberg plate of 1918 February 4, as a star of $8^{m}.5$. On Harvard plates, the nova is [$10^{m}.8$ on 1917 December 22, $5^{m}.7$ on 1918 January 1 (*AN* **206** (1918) 57)).

Position:			
	07 24 20.60	– 06 34 23.85	(POSS)
	07 24 20.73	– 06 34 23.5	(5 outburst observations)
	222.930	+ 4.749	(G.C.)

Range: 5.6p (5.2:) – 18p LCT: A? t_3: ~ 23^d

Finding chart: M. Wolf, *AN* **206** (1918) 99.

Light curve: H. S. Leavitt, *HA* **84** (1920) 121; G. Cecchini, L. Gratton (1941) 105; C. Payne-Gaposchkin (1957) 9.

Spectroscopy: A. J. Cannon, *HC* 209 (1918) – descr; G. F. Paddock, *Lick Bull* 313 (1919) – descr, ident; W. S. Adams, A. H. Joy, *PASP* **30** (1918) 162, 193 – descr.

Identification: from Heidelberg plates B 4057/4058, taken 1918 February 17.

Classification: fast nova; maximum poorly covered by observations.

KT Mon **NA**

(N Mon 1942)

Discovered by A. N. Vyssotzky on a Harvard spectral plate taken 1943 January 2 (*AJ* **59** (1954) 199).

Position:			
	06 22 38.43	+ 05 28 16.3	(POSS, empty field)
	06 22 38.8	+ 05 28 13	(S. Gaposchkin, *AJ* **59** (1954) 199)
	205.100	– 3.299	(G.C.)

Range: 10.3p (9.8) – [21 LCT: A? Ba? t_3: 40^d

Finding chart: A. Sh. Khatisov (1971); incorrect identification and position.
Light curve: S. Gaposchkin, *AJ* **59** (1954) 199.
Spectroscopy: S. Gaposchkin, *AJ* **59** (1954) 199.
Identification: from Harvard plates IR 6448, taken 1943 February 22/23, and IR 6466, taken 1943 March 8/9. No POSS counterpart.
Classification: moderately fast nova.

V616 Mon XND

(N Mon 1975, A0620-00, Mon X-1)

Discovered as an X-ray flaring source by M. Elvis, C. G. Page, K. A. Pounds, M. J. Ricketts and M. J. L. Turner with the Ariel-5 satellite in August 1975 (*Nature* **257** (1975) 656).

Position: 06 20 11.14 − 00 19 10.7 (POSS)
06 20 11.148 − 00 19 10.9 (L. J. Eachus, E. L. Wright, W. Liller, *ApJ* **203** (1976) L17)
209.956 − 6.540 (G.C.)

Range: 11.2p – 20.2B LCT: pec t_3: ?

Finding chart: F. Boley, R. Wolfson, H. Bradt, R. Doxsey, G. Jernigan, W. A. Hiltner, *ApJ* **203** (1976) L13; H. W. Duerbeck, W. C. Seitter, *SuW* **14** (1976) 347.
Light curve: R. F. Webbink, 'A provisional optical light curve of the X-ray recurrent nova V616 Mon' (1978), Dept. Astr. Univ. Illinois, Urbana, Ill.; C. Lloyd, R. Noble, M. V. Penston, *MN* **179** (1977) 675; L. J. Eachus, E. L. Wright, W. Liller, *ApJ* **203** (1976) L17 – outburst of 1917.
Spectroscopy: J. A. J. Whelan, M. J. Ward, D. A. Allen, J. Danziger, R. A. E. Fosbury, G. P. Murdin, M. V. Penston, E. J. Wampler, B. L. Webster, *MN* **180** (1977) 657.
X-ray observations: R. Doxsey, G. Jernigan, D. Hearn, H. Bradt, J. Bult, G. W. Clark, J. Delvaille, A. Epstein, P. C. Joss, T. Matilsky, W. Mayer, J. McClintock, S. Rappaport, J. Richardson, H. Schnopper, *ApJ* **203** (1976) L9; R. E. Griffiths, M. J. Ricketts, B. A. Cooke, *MN* **177** (1976) 429; L. J. Kaluzienski, S. S. Holt, E. A. Boldt, P. J. Serlemitsos, *ApJ* **212** (1977) 203.
Radio observations: F. N. Owen, T. J. Balonek, J. Dickey, Y. Terzian, S. T. Gottesman, *ApJ* **203** (1976) L15.
Duplicity: J. E. McClintock, R. A. Remillard, *ApJ* **308** (1986) 110.
Identification: from Duerbeck and Seitter's finding chart.
Classification: recurrent soft X-ray flaring object; outbursts occurred in 1917, 1975. Blue, nearly featureless continuum during outburst.

GQ Mus **NA**

(N Mus 1983)

Discovered by W. Liller, Viña del Mar, Chile, 1983 January 18, as a star of $7^{m}_{.}2$ (*IAU Circ* 3764).

Position:	11 49 34.99	− 66 55 39.0	(SRC)
	11 49 35.13	− 66 55 39.3	(GPO plate, May 1986, decline)
	11 49 35.11	− 66 55 38.3	(I. N. Nikoloff, J. Johnson, *IAU Circ* 3766 (1983))
	297.212	− 4.996	(G.C.)

Range: 7.2v − 22:j LCT: (Ao?) t_3: 45^{d}

Finding chart:	J. Krautter, K. Beuermann, C. Leitherer, E. Oliva, A. F. M. Moorwood, E. Deul, W. Wargau, G. Klare, L. Kohoutek, J. van Paradijs, B. Wolf, *AsAp* **137** (1984) 307.
Light curve:	F. M. Bateson, A. W. Dodson, *NZAS Publ* **11** (1983) 11.
Spectroscopy:	B. S. Shylaya, *Obs* **103** (1983) 203 – spectrophotometry; J. A. de Freitas Pacheco, S. J. Codina, *MN* **214** (1985) 481.
UV observations:	J. Krautter *et al.* (1984).
IR observations:	J. Krautter *et al.* (1984); P. A. Whitelock, B. S. Carter, M. W. Feast, I. S. Glass, D. Laney, J. W. Menzies, J. Walsh, P. M. Williams, *MN* **221** (1984) 421; H. Dinerstein, *AJ* **92** (1986) 1381 – IRAS observations.
X-ray observations:	H. Ögelman, K. Beuermann, J. Krautter, *ApJ* **287** (1984) L31; H. Ögelman, K. Beuermann, J. Krautter, ESA workshop 'Recent Results on Cataclysmic Variables', *ESA SP-236* (1986) 177.
Identification:	from the finding chart by Krautter *et al.*
Classification:	moderately fast nova, observed over a large wavelength range.

IL Nor

(N Nor 1893, N Nor No. 1, R Nor (renamed!), HV 25, HD 137677)

Discovered by W. Fleming on a Harvard spectral plate taken 1893 October 26. The star was [14^{m} on 1893 June 21, 7^{m} on July 10 (*AN* **134** (1893) 59).

Position:	(1)	15 25 45.56	− 50 24 42.05	(SRC; star 18^{m})
	(2)	15 25 45.62	− 50 24 45.15	(SRC; star 20^{m})
	(3)	15 25 45.26	− 50 24 40.4	(SRC, star 20^{m})
		15 25 47	− 50 24 37	(E. C. Pickering, *AN* **134** (1893) 181)
		326.835	+ 4.810	(G.C.)

Range: **7.0**p – 18j? LCT: B? t_3: 108^d

Finding chart: W. Fleming, *HA* **26** (1897) 257.
Light curve: W. Fleming, *HA* **26** (1897) 257; A. D. Walker, *HA* **84** (1923) 189; G. Cecchini, L. Gratton (1941) 48, 49.
Spectroscopy: W. W. Campbell, *PASP* **6** (1894) 102 – descr; A. J. Cannon, *HA* **76** (1916) 19 – descr.
Identification: from Harvard plates X 5784, taken 1894 August 1, X 5792, 1894 August 2, and X 5802, 1894 August 3; the nova coincides with a blend of three stars, which are listed above.
Classification: moderately fast nova; maximum poorly covered.

IM Nor N:

(N Nor 1920, N Nor No. 2, HV 3532)

Discovered by I. E. Woods on a Harvard plate taken 1920 July 7 as a 9^m star (*HB* 734 (1921), *AN* **213** (1921) 47).

Position: (1) 15 35 42.26 – 52 09 39.8 (SRC, star 22^m)
(2) 15 35 42.64 – 52 09 38.8 (SRC, star 22^m)
15 35 42.39 – 52 09 37.9 (Harvard plate MF 6504)
327.097 + 2.485 (G.C.)

Range: 9.0p – 22.0j LCT: C? t_3: ?

Finding chart: S. Wyckoff, P. A. Wehinger, *PASP* **91** (1979) 173.
Light curve: J. L. Elliot, W. Liller, *ApJ* **175** (1972) L69.
Identification: from Harvard plate MF 6504, taken 1920 July 27. Wyckoff and Wehinger's position is incorrect.
Classification: nova without spectroscopic verification.

QX Nor XND

(4U1608-52, MX1608-52)

Recurrent flaring (or transient) X-ray source. Optical identification by J. E. Grindlay and W. Liller (*ApJ* **270** (1978) L127).

Position: 16 08 52.2 – 52 17 43 (J. E. Grindlay and W. Liller (*ApJ* **270** (1878) L127)
330.926 – 0.850 (G.C.)

Range: 18.2ir – [20ir LCT: ? t_3: ?

Finding chart: J. E. Grindlay and W. Liller, *ApJ* **270** (1978) L127.
Identification: from Grindlay and Liller's finding chart, empty field.
Classification: GCVS classification: XND.

V341 Nor N?

(N Nor 1983)

Discovered by W. Liller, Viña del Mar, Chile, 1983 September 19 as a star of $9^{m}.4$; it was [$11^{m}.5$ on 1983 September 10 (*IAU Circ* 3869).

Position:			
(1)	16 09 50.68	− 53 11 31.95	(SRC, star 23^{m})
(2)	16 09 50.81	− 53 11 28.8	(SRC, star 21^{m})
(3)	16 09 50.21	− 53 11 33.3	(SRC, star 17^{m})
	16 09 51.0	− 53 11 32	(W. Liller, *IAU Circ* 3874 (1983))
	330.423	− 1.607	(G.C.)

Range: 9.4v – [17j LCT: ? t_3: ?

Identification: from Liller's position; not verified by additional observations. Three stars in the vicinity of Liller's position are listed.
Classification: poorly known object. Amplitude suggests nova.

N Nor 1985/1 M:

Discovered by W. Liller, Viña del Mar, Chile, 1985 January 26, as a star of $10^{m}.5$ (*IAU Circ* 4030).

Position:		
15 36 48.94	− 51 03 14.7	(SRC)
15 36 49.07	− 51 03 15.1	(R. H. McNaught, *IAU Circ* 4035 (1985))
327.893	+ 3.274	(G.C.)

Range: 10.5v – [16.0 LCT: ? t_3: –

Identification: from McNaught's precise position.
Classification: spectral type M2-M5; semiregular or Mira variable (*IAU Circ* 4033 (1985)).

N Nor 1985/2 M or ZAND

(Liller's variable; in the vicinity of NSV 7429)

Discovered by W. Liller, Viña del Mar, Chile, 1985 May 28 as a star of 9^{m} (red) (*IAU Circ* 4075).

taken 1919 November 15, and MC 16377, taken 1919 November 15; blue star on POSS.

Classification: slow nova; the well-observed light curve has two maxima of nearly equal brightness.

V906 Oph **NA**

(N Oph 1952)

Discovered by B. Iriate on a Tonantzintla objective prism plate taken 1952 August 22, as an 11^m star (*IAU Circ* 1371).

Position:	17 23 28	– 21 50 06	(A. V. Soloviev, *ATs* 130 (1952) 3)
	3.654	+ 7.430	(G.C.)

Range: 8.4 – ? LCT: ? t_3: 25^d

Finding chart: G. Rüdiger, *MVS* **2** (1965) 166; W. Wenzel, *IBVS* 2585 (1984); A. Sh. Khatisov (1971) – identification doubtful.

Light curve: G. de Vaucouleurs, *HAC* 1187 (1952), *IAU Circ* 1371 (1952); W. Wenzel, *IBVS* 2585 (1984).

Spectroscopy: W. A. Hiltner, *ApJ* **120** (1954) 596 – phot.

Identification: unclear. The finding charts by Rüdiger and Wenzel are not accurate enough. Khatisov incorrectly identifies a star of 12^m with a late type spectrum as the exnova.

Classification: fast nova, poorly known.

V908 Oph **N**

(N Oph 1954)

Discovered by V. Blanco, Warner and Swasey Observatory, as an emission-line object on an objective prism plate taken 1954 July 2 (*IAU Circ* 1461, *HAC* 1257, *NBl AZ* **8**, 33).

Position:	17 24 42	– 27 43	(V. Blanco, *IAU Circ* 1461 (1954))
	358.893	+ 3.932	(G.C.)

Range: 9 (7.5) – ? LCT: ? t_3: ?

Spectroscopy: K. Seyfert, K. Yoss, *HAC* 1260 (1954).

Identification: not possible. Field map in Appendix.

Classification: poorly known nova.

V972 Oph **NB**

(N Oph 1957)

Discovered by G. Haro on Tonantzintla plates taken 1957 September 18, as a star of $9^{m}.8$ (*HAC* 1405).

Position:	17 31 34.61	− 28 08 38.7	(SRC)
	17 31 39	− 28 08	(G. Haro, *HAC* 1405 (1957))
	359.374	+ 2.430	(G.C.)

Range: **8.0**p − 17.0j LCT: D t_3: 176^d

Finding chart:	E. de la Rosa, *TTB* **18** (1959).
Light curve:	E. de la Rosa, *TTB* **18** (1959).
Spectroscopy:	E. de la Rosa, *TTB* **18** (1959) – phot, descr; H. W. Duerbeck, W. C. Seitter, *ApSS* **131** (1987) 467 – minimum, descr.
Identification:	on POSS plate in outburst; comparison with SERC equatorial survey.
Classification:	slow nova with pre-maximum halt of at least 50^d

V1012 Oph **N**

(N Oph 1961, SVS 1308)

Discovered by S. P. Apriamashvili, Abastumani Observatory, on an objective prism plate taken 1961 April 20 (*ATs* 221 (1961) 1, *IAU Circ* 1758).

Position:	17 38 31.92	− 23 22 06.0	(SRC)
	17 38 31.98	− 23 22 06.0	(O. D. Dokuchaeva, Yu. N. Efremov, *ATs* 221 (1961) 2)
	17 38 32.06	− 23 22 08.5	(A. Sh. Khatisov (1971))
	4.250	+ 3.677	(G.C.)

Range: 13.8p − 22j LCT: ? t_3: ?

Finding chart:	O. D. Dokuchaeva, Yu. N. Efremov, *ATs* 221 (1961) 2; A. Sh. Khatisov (1971).
Light curve:	S. P. Apriamashvili, O. D. Dokuchaeva, Yu. N. Efremov, *ATs* 221 (1961).
Spectroscopy:	S. P. Apriamashvili, *ATs* 221 (1961) 1.
Identification:	the position published by O. D. Dokuchaeva and Yu. N. Efremov coincides with a star of 22^{m} on the SERC J plate. The authors state that the prenova is fainter than the POSS limit ([$20^{m}.8$). Khatisov's position yields an empty field.
Classification:	poorly known nova.

V1195 Oph **M:**

(Plaut No. 71)

Discovered by L. Plaut, Leiden Observatory. Maxima occurred 1956 June 12 and 1959 April 19 (*BAN Suppl* **3**, 1 (1968)).

Position: 16 57 24 − 20 49 06 (L. Plaut, *BAN Suppl* **3**, 1 (1968))
0.969 + 12.988 (G.C.)

Range: 16.3p – [20.1p LCT: ? t_3: ?

Light curve: L. Plaut, *BAN Suppl* **3**, 1 (1968).
Identification: not possible because of lack of precise position or finding chart. Field map in Appendix.
Classification: Plaut classifies the variable as recurrent nova or long period variable. The second classification is more probable because of the red colour of the object. The GCVS4 classifies the variable as NR:

V1235 Oph **N: or UV**

(Plaut No. 111)

Discovered by L. Plaut, Leiden Observatory. The star is visible on 1959 August 29 only (*BAN Suppl* **3**, 1 (1968)).

Position: 16 58 52 − 20 16 42 (L. Plaut, *BAN Suppl* **3**, 1 (1968))
1.624 + 13.031 (G.C.)

Range: 17.0p – [20.8p LCT: ? t_3: ?

Light curve: L. Plaut, *BAN Suppl* **3**, 1 (1968).
Identification: not made because of lack of precise position or finding chart. Field map in Appendix.
Classification: Plaut classifies the variable as nova or flare star; the colour index is − 0.5. The GCVS4 lists the variable as N: or UV.

V1548 Oph **N?**

(Plaut No. 429)

Discovered by L. Plaut, Leiden Observatory. The star is visible only on two plates taken 1959 June 27 (*BAN Suppl* **3**, 1 (1968)).

Position: 17 08 33 − 16 16 42 (L. Plaut, *BAN Suppl* **3**, 1 (1968))
6.343 + 13.455 (G.C.)

Range: 13.7p – [20 LCT: ? t_3: ?

Light curve: L. Plaut, *BAN Suppl* **3**, 1 (1968).
Identification: not attempted because of lack of precise position or finding chart. Field map in Appendix.
Classification: The GCVS4 lists the star as N:, a classification which is in agreement with the observed amplitude.

V2024 Oph N

(N Oph 1967)

Discovered by C. B. Stephenson on objective prism plates taken at Cerro Tololo, 1967 July 7, when the nova was about $1^{m}.5$ below maximum (*IBVS* 323 (1968)).

Position: 17 39 18 − 24 58 (C. B. Stephenson, *IBVS* 323 (1968))
2.986 + 2.681 (G.C.)

Range: 11.0v (9.5) – [18 LCT: ? t_3: ?

Identification: prenova certainly fainter than 18^{m}, not identifiable (C. B. Stephenson, *IBVS* 323 (1968)). Field map in Appendix.
Classification: poorly known nova.

V2104 Oph N

(N Oph 1976)

Discovered by T. Kuwano, Japan, 1976 September 23, as a star of $8^{m}.8$ (*IAU Circ* 2994).

Position: 18 01 05.37 + 11 47 47.3 (POSS)
18 01 05.19 + 11 47 46.8 (2 outburst observations)
38.251 + 15.928 (G.C.)

Range: 8.8v (5.3) – 20.5p LCT: ? t_3: ?

Light curve: H. Huth, *IBVS* 1205 (1976).
Spectroscopy: P. Pesch, *IAU Circ* 2996 (1976).
Identification: Pesch notes that the prenova is not visible on the POSS red plate. Our identification is based on the precise positions determined at maximum light.
Classification: poorly known nova.

V2107 Oph **XND**

(N Oph 1977, H1705-25)

X-ray nova, discovered with the HEAO-1 satellite, optically identified by R. E. Griffiths *et al.* (1978).

Position:	17 05 10.05	− 25 01 41.4	(SRC)
	17 05 10.4	− 25 01 38	(R. E. Griffiths *et al.* 1978)
	358.586	+ 9.058	(G.C.)

Range: 16.6p – 21.0p LCT: ? t_3: ?

Finding chart: R. E. Griffiths, H. Bradt, R. Doxsey, H. Friedman, H. Gurski, M. Johnson, A. Longmore, D. F. Malin, P. Murdin, D. A. Schwartz, J. Schwarz, *ApJ* **221** (1978) L63.
Spectroscopy: R. E. Griffiths *et al.* (1978)
X-ray observations: M. G. Watson, M. J. Ricketts, R. E. Griffiths, *ApJ* **221** (1978) L69; B. A. Cooke, A. M. Levine, F. L. Lang, F. A. Primini, W. H. G. Lewin, *ApJ* **285** (1984) 258.
IR observations: R. M. Catchpole, I. S. Glass, G. Roberts, J. Spencer Jones, P. Whitelock, *SAAO Circ* **9** (1985) 1.
Identification: from the finding chart of Griffiths *et al.*.
Classification: X-ray nova, similar to V616 Mon.

V2109 Oph **N**

(N Oph 1969)

Discovered by D. J. MacConnell on an objective prism plate taken with the Cerro Tololo Schmidt telescope, 1969 June 21 (*IBVS* 1340 (1977)).

Position:	17 21 12.08	− 24 34 08.05	(SRC)
	17 21 12.14	− 24 34 08.0	(D. J. MacConnell, *IBVS* 1340 (1977)
	1.074	+ 6.342	(G.C.)

Range: 10.8r – 18j LCT: ? t_3: ?

Identification: from D. J. MacConnell's precise position; confirmation is necessary.
Classification: poorly known nova.

V2110 Oph **NB or ZAND**

(AS 239, Hen 1465)

Discovered as an emission line object of 12^m by P. W. Merrill and C. G. Burwell (*ApJ* **112** (1950) 72); in a similar survey around 1950, it was found by K. G. Henize as an object of $13^m.5$. On UK Schmidt plates, it is $18^m.5$ in 1976, $19^m.5$ in 1978 (D. A. Allen, *IBVS* 1399 (1978)).

Position:	17 40 31.60	− 22 44 18.5	(SRC)
	17 40 31	− 22 44 16	(D. A. Allen, *IBVS* 1399 (1978))
	5.029	+ 3.620	(G.C.)

Range: 12.0p – 20j LCT: ? t_3: slow?

Finding chart:	D. A. Allen, *IBVS* 1399 (1978).
Light curve:	D. A. Allen, *IBVS* 1399 (1978).
IR photometry:	M. W. Feast, I. S. Glass, *Obs* **100** (1980) 208.
Spectroscopy:	D. A. Allen, *IBVS* 1399 (1978).
Identification:	from Allen's finding chart.
Classification:	slow nova or symbiotic object; poorly known.

N Oph 1893 **N?**

(NSV 07542)

Reported missing in 1926 by L. Fillippoff, Observatoire d'Alger, as star No. 8 on the Carte du Ciel plate San Fernando $-5°\,16^h12^m$, taken 1893 July 1 (*AN* **228** (1926) 329).

Position:	16 10 51.67	− 05 11 04.45	(POSS)
	16 10 51.53	− 05 11 09.3	(L. Fillippoff, *AN* **228** (1926) 329)
	7.160	+ 31.415	(G.C.).

Range: 11.5p – 21j? LCT: ? t_3: ?

Finding chart:	L. Fillippoff, *AN* **228** (1926) 329.
Identification:	from Fillippoff's accurate position. The faint object listed above is 5″ from the Carte du Ciel position. Other stars are more than 15″ distant. Tentative identification.
Classification:	the nature of the object is not clear: nova or asteroid.

N Oph 1938 N:

Discovered by H. Sawyer-Hogg and A. Wehlau, David Dunlap Observatory, as a star of $15^{m}.6$ in the globular cluster NGC 6402 (M 14) (*JRAS Can* **58** (1963) 163, *AJ* **69** (1964) 141).

Position: 17 35 00.400 − 03 12 54.5 (M. Shara *et al.* (1986))
21.329 + 14.799 (G.C.)

Range: 15.6p (10?) – 21v LCT: ? t_3: ?

Finding chart: M. M. Shara, A. F. J. Moffat, M. Potter, H. Sawyer Hogg, A. Wehlau, *ApJ* **311** (1986) 796. Finding chart in Appendix.
Light curve: H. Sawyer Hogg, A. Wehlau, *JRAS Can* **58** (1963) 163.
Identification: the atlas is not well suited to show the object. The reader is referred to the published finding chart by M. M. Shara, A. F. J. Moffat, M. Potter, H. Sawyer-Hogg, A. Wehlau, *ApJ* **311** (1986) 796.
Classification: nova without spectroscopic observation. The early stages of the outburst were not observed. Probably a member of NGC 6402.

FU Ori FU

(N Ori 1939, 1.1939 Ori)

Discovered as a nova-like object by. A. A. Wachmann, Hamburger Sternwarte, on a plate taken 1939 January 18 (*IAU Circ* 738).

Position: 05 42 37.972 + 09 03 02.54 (A. A. Wachmann, *ZsAp* **35** (1954) 74; A. Sh. Khatisov (1971))
197.109 − 10.251 (G.C.)

Range: 16.0 p (before outburst) – 9.6p LCT: pec t_3: –

Finding chart: A. A. Wachmann, *ZsAp* **35** (1945) 74; A. Sh. Khatisov (1971); G. H. Herbig, *Vistas* **8** (1966) 109.
Light curve: A. A. Wachmann, *ZsAp* **35** (1954) 74; D. Hoffleit, *HB* 911 (1939); G. H. Herbig, *ApJ* **217** (1977) 693; G. H. Herbig, *Vistas* **8** (1966) 109; E. A. Kolotilov, P. P. Petrov, *Pisma AZh* **11** (1985) 846 = *Sov Astr Letters* **11** (1985) 358.
Spectroscopy: P. Wellmann, *ZsAp* **29** (1951) 292 – ident, rv, line strengths; G. H. Herbig, *Vistas* **8** (1966) 109 – phot; L. V. Kuhi, *AsAp Suppl* **15** (1974) 47 – spectrophotometry.
Identification: from published finding charts.
Classification: prototype of pre-main sequence stars of the FU class (GCVS4).

GR Ori **N:**

(N Ori 1916, 2.1916 Ori)

Discovered by H. Thiele, Hamburger Sternwarte, on plates taken 1916 January 30 (*AN* **202** (1916) 213).

Position:	05 19 00.20	+ 01 07 15.9	(H. Thiele, *AN* **202** (1919) 213)
	201.172	− 19.310	(G.C.)

Range: 11.5p − [21p LCT: ? t_3: −

Finding chart: G. H. Herbig, *PASP* **70** (1958) 606.
Light curve: H. Thiele, *AN* **202** (1916) 213.
Identification: from Herbig's finding chart; empty field on POSS.
Classification: poorly known object. Amplitude suggests nova.

V529 Ori **N??**

(48Hev Ori, N Ori 1667 (!))

Discovered by J. Hevelius, Danzig/Gdansk, 1678 March 28, during a lunar occultation (Machina Coelestis II, Gedanum 1679, p. 804, 814).

Position:	05 55 24($\pm$8.5)	+ 20 15 12($\pm$30)	(W. B. Ashworth, *QJRAS* **22** (1981) 22)
	188.94	− 1.95	(G.C.)

Range: 6v − 20.5p? LCT: ? t_3: ?

Identification: The position was determined by W. B. Ashworth (*QJRAS* **22** (1981) 22), who reduced Hevelius' original observations. An identification is not possible without additional information. A blue star of 20^m, the NW component of a close visual pair, is the best candidate in the error box given by Ashworth:

(1)	05 55 18.53	+ 20 15 06.7	NW component
(2)	05 55 18.69	+ 20 15 02.1	SE component
(3)	05 55 30.03	+ 20 15 08.75	bright star

The identification and position given by A. Sh. Khatisov (1971) are certainly incorrect.

Classification: amplitude suggests nova. The reality of the object is, however, not firmly established. The GCVS classification, NR:, is not correct in view of Ashworth's study.

BD Pav UG

(N Pav 1934, HV 10031)

Discovered by C. D. Boyd on Harvard plates. Maximum light occurred 1934 September 7 (*HA* **90** (1939) 248. A second outburst was observed in 1985.

Position:	18 38 54.66	− 57 33 43.0	(SRC)
	18 38 55	− 57 34 36	(C. D. Boyd, *HA* **90** (1939) 248)
	338.182	− 21.558	(G.C.)

Range: 12.4p – 16.5j LCT: DN t_3: ?

Finding chart: S. Wyckoff, P. A. Wehinger (1971)
Light curve: C. D. Boyd, *HA* **90** (1939) 248.
Duplicity: periodic light variations with an amplitude of $0^{m}.5$ and P = 0.1793015 d at minimum (H. Barwig, R. Schoembs, *IBVS* 2031 (1981), *AsAp* **124** (1983) 287).
Identification: from Harvard plate MF 19815, taken 1934 September 7/8.
Classification: a second outburst during the summer of 1985 (H. Barwig, private communication) and the small amplitude indicate that the object is a dwarf nova with long intervals between outbursts.

V Per N

(N Per 1887, N Per No. 1, BD +56°406[a], HD 12244, HV 6)

Discovered by W. Fleming on Harvard plates taken in 1887. The observed maximum magnitude occurred 1887 December 19 (*HC* 4 (1890), *AN* **126** (1890) 163).

Position:	01 58 29.305	+ 56 29 36.8	(POSS)
	01 58 29.47	+ 56 29 33.7	(K. Graff, *AN* **198** (1914) 145)
	132.521	− 4.820	(G.C.)

Range: 9.2p (4?) – 18.5p LCT: ? t_3: ?

Finding chart: E. Hartwig, *Bamb Ver* **1** (1913) 92.
Light curve: H. Leavitt, *HA* **84** (1920) 121.
Spectroscopy: A. J. Cannon, *HA* **76** (1916) 19 – descr; D. B. McLaughlin, *PASP* **58** (1946) 218 – descr; H. W. Duerbeck, W. C. Seitter (unpublished) – minimum spectrum.
Identification: rediscovered by its spectral appearance at minimum.
Classification: nova; maximum not covered by observations.

SZ Per

N??

(143.1908 Per, BD +33°715)

Reported missing by W. Luther (*AN* **179** (1908) 175). BD observations of 1853 November 30 and 1865 October 30 shows a star of $9^{m}_{.}5$, which is also seen on a Carte du Ciel plate, Potsdam, 1894 November 30 (10^{m}), but not on 1899 October 15 ([$10^{m}_{.}5$).

Position:			
(1)	03 43 54.20	+34 11 15.1	(POSS)
(2)	03 43 55.925	+34 11 25.5	(POSS)
(3)	03 43 55.325	+34 11 47.0	(POSS)
(4)	03 43 54.165	+34 11 57.9	(POSS)
(5)	03 43 53.93	+34 12 02.2	(POSS)
(6)	03 43 58.26	+34 08 57.1	(POSS, Zinner's (1939) star)
	03 43 54.7	+34 11 06	(BD)
	159.458	−15.792	(G. C. of BD position)

Range: 9.5v – ? LCT: ? t_3: ?

Light curve: E. Zinner, *AN* **267** (1939) 61.

Identification: stars Nos. 1–5 near the BD position. Zinner's identification (star No. 6) is doubtful (K. Himpel, *BZ* **24** (1942) 40). Himpel ignores the sightings on the Potsdam plates (which may refer to another star), and suspects that the BD observations indicate a nova. The time interval between the two BD observations makes this conjecture extremely unlikely.

Classification: doubtful object, probably not a nova.

UW Per

UG or N:

(21.1912 Per)

Discovered by C. R. d'Esterre, Tatsfield, England, on plates taken in 1912 January. Maximum light occurred 1912 January 12 (*AN* **191** (1912) 63).

Position:		
02 09 00.415	+56 51 16.4	(POSS)
02 08 03	+56 54 30	(C. R. d'Esterre, *AN* **191** (1912) 63)
133.806	− 4.062	(G.C.)

Range: 13.5p – 18.0p? LCT: ? t_3: ?

Finding chart: C. R. d'Esterre, *AN* **191** (1912) 63.

Light curve: C. R. d'Esterre, *AN* **191** (1912) 63; E. Zinner, *Bamb Ver* **1** (1932) 594; Yu. N. Efremov, *PZv* **14** (1962) 258.

Identification:	from d'Esterre's finding chart; tentative identification. The candidate is a blend of three stars; the position refers to the brightest component. d'Esterre's position seems to be erroneous.
Classification:	the fairly rapid decline to minimum brightness which is indicated by d'Esterre's observations and the amplitude indicate a dwarf nova. This is supported by three additional brightenings between 1915 and 1922 (E. Zinner, *Bamb Ver* **1** (1932) 594).

GK Per NA

(N Per 1901, N Per No. 2, 3.1901 Per, BD +43°740[a], HD 21629)

Discovered by T. D. Anderson, Edinburgh, 1901 February 21, as a star of $2^{m}.7$. It was invisible during the preceding nights (*AN* **154** (1901) 363).

Position:	03 27 47.36	+43 44 05.1	(3 new observations)
	03 27 47.61	+43 44 05.1	(28 outburst observations)
	150.955	−10.104	(G.C.)

Range: **0.2**v − 11.8 ... 14.0 LCT: Ao t_3: 13^d

Finding chart:	M. Humason (1938); A. Sh. Khatisov (1971); G. Williams (1983).
Light curve:	L. Campbell, *HA* **48** (1903) 39; H. Leavitt, *HA* **84** (1920) 121; G. Cecchini, L. Gratton (1941) 64–66; C. Bertaud (1945) 33; C. Payne-Gaposchkin (1957) 9, 140; F. Sabbadin, A. Bianchini, *AsAp Suppl* **54** (1983) 393; pre-outburst: E. L. Robinson, *AJ* **80** (1975) 575; A. Bianchini, F. Sabbadin, E. Hamzaoglu, *AsAp* **106** (1982) 176, A. Bianchini, F. Sabbadin, G. C. Favero, I. Dalmeri, *AsAp* **160** (1986) 367 – optical outbursts at minimum.
Spectroscopy:	G. E. Hale, *ApJ* **13** (1901) 173, 238 – phot; H. C. Vogel, *ApJ* **13** (1901) 217 – trac; W. S. Adams, *ApJ* **14** (1901) 158 – ident; W. W. Campbell, W. H. Wright, *ApJ* **14** (1901) 269 – phot, trac, ident; A. J. Cannon, *HA* **56** (1912) 41, *HA* **76** (1916) 19 – phot, ident, descr; F. J. M. Stratton, *ASPO Camb* **4**, 2 (1936) 73 – phot, ident, rv; D. B. McLaughlin, *Michigan Publ* **9** (1949) 13 – phot, ident, rv, line intensities; R. P. Kraft, *ApJ* **139** (1964) 457 – minimum spectrum, phot; G. Williams (1983) – minimum spectrum, trac.
Nebular shell:	F. G. Pease, G. W. Ritchey, *PASP* **29** (1917) 256; E. E. Barnard, *AJ* **30** (1917) 86; G. W. Ritchey, *PASP* **30** (1918) 163; W. C. Seitter, H. W. Duerbeck, in 'RS Oph and the recurrent nova phenomenon' (ed. M. F. Bode, Utrecht 1986); H. W. Duerbeck, *ApSS* (in press); S. P. Reynolds, R. A. Chevalier, *ApJ* **281** (1984) L33 – radio emission.

Illumination of surrounding nebulosity: M. Wolf, *AN* **156** (1901) 253; C. D. Perrine, *ApJ* **14** (1902) 249; A. Kopff, *Heidelberg Ver* **2** (1906) 14; P. Couderc, *AAp* **2** (1939) 271.

UV observations: L. Rosino, A. Bianchini, P. Rafanelli, *AsAp* **108** (1982) 243.

X-ray observations: F. A. Cordova, K. O. Mason, J. E. Nelson, *ApJ* **245** (1981) 609; A. R. King, M. J. Rickett, R. S. Warwick, *MN* **197** (1979) 77p; M. G. Watson, A. R. King, J. Osborne, *MN* **212** (1985) 917; A. Bianchini, F. Sabbadin, *IBVS* 2751 (1985).

Duplicity: spectroscopic binary, P = 1.996803 d (H. Ritter 1984; contains additional references); D. Crampton, A. P. Cowley, W. A. Fisher, *ApJ* **300** (1986) 788).

Identification: from published finding charts.

Classification: well-observed very fast nova with unusual characteristics: light echo after outburst, quite asymmetric nebular remnant with radio emission, longest observed orbital period among classical novae.

V400 Per — NA

(N Per 1974)

Discovered by N. Sanduleak, Warner and Swasey Observatory, on an objective prism plate taken 1974 November 9 as an 11^m star (*IAU Circ* 2716).

Position:			
	03 04 12.435	+46 56 08.95	(POSS)
	03 04 12.50	+46 56 09.8	(2 outburst observations)
	145.663	− 9.644	(G.C.)

Range: 7.8p – 20p LCT: Ba t_3: 43^d

Finding chart: N. Sanduleak, *IBVS* 959 (1975); A. U. Landolt, *PASP* **87** (1975) 407; A. Sh. Khatisov, *ATs* 856 (1974) 2.

Light curve: W. Wenzel, *IBVS* 947 (1974), *MVS* **6** (1975) 201; N. N. Narizhnaya, N. N. Kiselev, *ATs* 888 (1975) 1; L. Rosino, *ApSS* **55** (1978) 383.

Spectroscopy: L. Rosino, *ApSS* **55** (1978) 383 – trac, ident, rv; E. J. Weiler, J. D. R. Bahng, *MN* **174** (1976) 563 – spectrophotometry.

Identification: from Sanduleak's finding chart.

Classification: moderately fast nova.

RR Pic — NB

(N Pic 1925)

Discovered by R. Watson, West Beaufort, South Africa, 1925 May 25, as a star of $2^m.4$. The nova was 12.8p on 1925 February 18, 3^m on 1925 April 13. Maximum was reached on 1925 on June 9 (*HB* 820 (1925); *PA* **33** (1925) 395).

Position:	06 35 09.80	− 62 35 49.3	(H. W. Duerbeck, M. Geffert, *IBVS* 2260 (1983))
	06 35 09.97	− 62 35 49.85	(2 outburst observations)
	272.355	− 25.672	(G.C.)

Range: **1.0**v – 11.9p LCT: D t_3: 150^d

Finding chart: S. Wyckoff, P. A. Wehinger (1978).

Light curve: H. Spencer Jones, *Cape Ann* **10** (1931) 9; S. Gaposchkin, *HA* 115 (1952) 111; L. Campbell, *HB* 835 (1929), *HB* 890 (1932); G. Cecchini, L. Gratton (1941) 129; C. Bertaud (1945) 96; C. Payne-Gaposchkin (1957) 15, 143; C. Payne-Gaposchkin, D. H. Menzel, *HC* 428 (1938) – continuum magnitudes.

Spectroscopy: H. Spencer Jones, *Cape Ann* **10** (1931) – phot, ident, rv; H. Spencer Jones, J. Lunt, *MN* **86** (1926) 498 – phot, ident, rv; W. H. Wright, *PASP* **37** (1925) 235, *PASP* **38** (1926) 233 – phot; C. Bertaud (1945) 96 – rv; C. Payne-Gaposchkin (1957) 142 – descr.

UV spectroscopy: H. W. Duerbeck, G. Klare, J. Krautter, B. Wolf, W. C. Seitter, W. Wargau, Proc. 2nd IUE Conf., *ESA SP 157* (1980) 91.

Nebular shell: R. E. Williams, J. S. Gallagher, *ApJ* **228** (1979) 482; W. C. Seitter, H. W. Duerbeck, *AG Mitt* **50** (1980) 70; W. C. Seitter, ESO Workshop on 'Production and Distribution of C,N,O Elements' (1985) 253.

Duplicity: spectroscopic binary with P = 0.1450255 d; light variations with the same period, eclipses shallow or absent; N. Vogt, *AsAp* **41** (1975) 15 – photometry; R. Haefner, K. Metz, *AsAp* **109** (1982) 171 – polarimetry; B. Warner, *MN* **219** (1986) 751 – photometry 1972 – 1984; S. Wyckoff, P. A. Wehinger, *IAU Coll 42* (= *Bamb Ver* **11**, 121 (1977) 201) – spectroscopic orbit; B. Warner, *MN* **195** (1981) 101 – rapid oscillations; M. Kubiak, *AA* **34** (1984) 331 – oscillations.

Identification: from Wyckoff and Wehinger's finding chart; the authors give an incorrect position.

Classification: well-investigated slow nova with extended pre-maximum halt (W. C. Seitter, *IAU Coll 4*, Budapest 1969, p. 277).

AS Psc **RN?**

(S 10828 in Triangulum)

Discovered by G. A. Richter on plates taken between 1963 September 14 and 17 with the Tautenburg Schmidt telescope. Maximum light occurred 1963 September 15. A second outburst was observed between 1980 August 8 and 14 (G. A. Richter, F. Börngen, *ApL* **21** (1981) 101; A. S. Sharov, *ATs* 1229 (1982), G. A. Richter, *ATs* 1262 (1983)).

Position: 01 25 20.24 + 30 59 38.5 (POSS)
01 25 24 + 30 59 (G. A. Richter, F. Börngen, *ApL* **21** (1981) 101)
132.099 − 30.974 (G.C.)

Range: **16.6**p − 21.5p? LCT: ? t_3: ?

Finding chart: G. A. Richter, F. Börngen, *ApL* **21** (1981) 101.
Light curve: G. A. Richter, F. Börngen, *ApL* **21** (1981) 101.
Identification: from Richter and Börngen's finding chart; tentative identification.
Classification: the very high UV and red excesses at outburst lead to the hypothesis that the object is a recurrent nova (G. A. Richter, *AN* **307** (1986) 221).

CP Pup **NA**

(N Pup 1942, 86.1942 Pup)

Discovered by B. H. Dawson, La Plata, 1942 November 9, as a star of $0^{m}.5$; it was already $1^{m}.1$ the previous night (*IAU Circ* 925).

Position: 08 09 52.04 − 35 12 04.35 (H. W. Duerbeck, M. Geffert, *IBVS* 2260 (1983))
08 09 52.12 − 35 12 03.7 (3 outburst observations)
252.926 − 0.835 (G.C.)

Range: **0.5**v − 15.0v LCT: A t_3: 8^d

Finding chart: E. Pettit, *PASP* **66** (1954) 142; J. Stein, J. Junkes, *Ric Astr* **1**, 10 (1945) 360.
Light curve: G. P. Kuiper, *ApJ* **97** (1943) 443; J. Stein, J. Junkes, *Ric Astr* **1**, 10 (1945) 360; S. Gaposchkin, *HB* 918 (1946); K. Tomida, M. Huruhata, *Tokyo Bull 2nd Ser* **11** (1948) 81; E. Pettit, *PASP* **61** (1949) 41, *PASP* **66** (1954) 142; D. J. K. O'Connell, *Pont Acad Sci Acta* **16** (1954) 49 = *Riv Repr* **8**; C. Payne-Gaposchkin (1957) 9, 146.
Spectroscopy: H. F. Weaver, *ApJ* **99** (1944) 280 – phot, ident, rv; R. F. Sanford, *ApJ* **102** (1945) 357 – phot, trac, ident, rv; R. F. Sanford, *PASP* **59** (1947) 334 – descr; D. B. McLaughlin, *ApJ* **118** (1953) 27 – coronal line; L. Gratton, *ApJ* **118** (1953) 586 – phot, rv.
H. Duerbeck, W. Seitter, R. Duemmler, *MN* (in press) – minimum spectrum, trac.
X-ray observations: R. F. Becker, L. Marshall, *ApJ* **244** (1981) L93.

Nebular shell:	W. C. Seitter, H. W. Duerbeck, *AG Mitt* **50** (1980) 70; R. E. Williams, *ApJ* **261** (1982) 170.
Duplicity:	spectroscopic binary with P = 0.061429 d (H. W. Duerbeck, W. C. Seitter, R. Duemmler, in press); A. Bianchini, M. Friedjung, F. Sabbadin, *IBVS* 2650 (1985); light variations with P = 0.06196 d (B. Warner, *MN* **217** (1985) 1p).
Identification:	from published finding charts.
Classification:	well-observed very fast nova.

DY Pup **NB:**

(N Pup 1902, HV 3600)

Discovered by I. E. Woods on Harvard plates. The object is first visible 1902 November 19 as 7^m star; it was [$10^m.3$ on 1902 October 24 (*HB* 760 (1921)).

Position:	08 11 42.57	– 26 24 48.6	(SRC)
	08 11 42.2	– 26 24 48	(H. Shapley, *HB* 760 (1921))
	245.823	+ 4.361	(G.C.)

Range: **7.0**p – 20p LCT: ? t_3: 160^d

Light curve:	H. Shapley, *HB* 760 (1921) – verbal description.
Identification:	from Harvard plates B 31244, taken 1902 January 15, and B 31506, taken 1903 April 21.
Classification:	no spectroscopic information available; amplitude and light curve suggest slow nova.

HS Pup **NA:**

(N Pup 1963 No. 2, BV 431)

Discovered by W. Strohmeier on plates of the Bamberg Southern Station, taken 1964 February 13 and later. The first rise occurred 1963 December 14/15, ($10^m.6$), and maximum light was reached 1963 December 23 ($8^m.0$) (*IBVS* 59 (1964)).

Position:	07 51 27.41	– 31 30 58.8	(SRC)
	07 51 29.6	– 31 31 21	(W. Strohmeier, *IBVS* 59 (1964))
	247.756	– 2.112	(G.C.)

Range: **8.0**p – 20.5p LCT: B? t_3: 65^d

Finding chart:	W. Strohmeier, *IBVS* 59 (1964); C. Hoffmeister, *Sterne* **40** (1964) 247.

Light curve:	W. Strohmeier, *IBVS* 59 (1964); H. Huth, C. Hoffmeister, *IBVS* 60 (1984).
Identification:	from a photograph supplied by R. Knigge, Bamberg.
Classification:	no spectroscopic observation is available; amplitude and light curve form suggest moderately fast nova. In obscured region of the Galaxy.

HZ Pup NA

(N Pup 1963 No. 1)

Discovered by C. Hoffmeister on Sonneberg plates. The nova appears first on a plate taken 1963 February 18 ($7^{m}.8$); it was [$12^{m}.0$ on 1962 December 28 (*IAU Circ* 1857 (1964), *IBVS* 45 (1964)).

Position:	08 01 20.215	−28 19 58.9	(SRC)
	08 01 19.7	−28 20 03	(C. Hoffmeister, *IBVS* 45 (1964))
	246.179	+ 1.385	(G.C.)

Range: **7.7**p – 17p LCT: Ca? t_3: 70^{d}

Finding chart:	C. Hoffmeister, *Sterne* **40** (1964) 247.
Light curve:	C. Hoffmeister, *AN* **288** (1965) 147.
Spectroscopy:	C. Hoffmeister, *Sterne* **40** (1964) 247.
Identification:	from Hoffmeister's finding chart and position.
Classification:	moderately fast nova with possible dust formation.

N Pup 1673 N??

(NVS 03846, Zi 816)

Discovered (or merely measured) by J. Richer on 1673 January 12 and 21 as star of 3^{m} with a mural quadrant set up in Cayenne (J. Richer, Recueil d'observations faites en plusieurs voyages ... Paris 1693).

Position:	07 57 21.24	−43 41 09.0	(SRC; blue star 20^{m})
	07 57 23	−43 40 52	(E. Zinner, *Bamb Ver* **2** (1926) 103; reduction of Richer's observation)
	258.809	− 7.424	(G.C.)

Range: 3v – 20p? LCT: ? t_3: ?

Identification:	from Richer's semi-precise position and the blue colour; confirmation by additional observations necessary.
Classification:	the reality of the object seems established. The amplitude suggests nova.

T Pyx **NR**

(N Pyx 1890, 1902; 32.1913 Pyx, HV 3348)

Discovered by H. Leavitt on plates of the Harvard Map (outburst of 1902) (*HC* 179 (1913), *AN* **197** (1914) 407).

Position:	09 02 37.14	− 32 10 47.3	(3 recent observations)
	257.207	+ 9.707	(G.C.)

Range: **6.5**p – 15.3p LCT: Dr t_3: 88^d

Finding chart:	M. Humason (1938); G. Williams (1983).
Light curve:	H. Leavitt, *HA* **84** (1920) 121; O. J. Eggen, D. S. Matheson, K. Serkowski, *Nature* **213** (1967) 216; G. Cecchini, L. Gratton (1941) 39; C. Payne-Gaposchkin (1957) 150.
Spectroscopy:	G. H. Herbig, *PASP* **57** (1945) 168 – ident, descr; A. H. Joy, *PASP* **57** (1954) 171 – phot, ident; G. Chincarini, L. Rosino, *IAU Coll. 4*, Budapest 1969, p. 261 – phot, trac, ident; R. Catchpole, *MN* **142** (1969) 119 – trac, ident, rv; G. Williams (1983) – minimum spectrum, trac.
UV observations:	A. Bruch, H. W. Duerbeck, W. Seitter, *AG Mitt* **52** (1982) 34.
Nebular shell:	W. C. Seitter, H. W. Duerbeck, *AG Mitt* **50** (1980) 70; R. E. Williams, *ApJ* **261** (1982) 170; W. C. Seitter, in 'RS Oph and the Recurrent Nova Phenomenon', ed. M. F. Bode, Utrecht 1986, p. 63.
Identification:	from Humason's finding chart.
Classification:	recurrent nova with slow development and variable surrounding nebulosity; outbursts in 1890, 1902, 1920, 1944, 1966.

SS Sge **ZAND or NC**

(11.1926 Sge, SVS 81)

Discovered by S. Beljawski on Simeis plates; a brightness increase between 1916 June and August was noted (*AN* **227** (1926) 423).

Position:	19 36 52.83	+ 16 35 44.25	(POSS)
	19 36 53	+ 16 35 45	(S. Beljawski, *AN* **227** (1926) 423)
	53.206	− 2.581	(G.C.)

Range: **11.8**p – 16.3p LCT: E? t_3: ?

Finding chart: S. Beljawski, *AN* **227** (1926) 423.

Light curve: K. Himpel, *BZ* **25** (1943) 106.
Identification: from Harvard plate MC 13030, taken 1917 July 22.
Classification: RT Ser-type nova according to K. Himpel (*BZ* **25** (1943) 106), nova-like variable according to C. Payne-Gaposchkin (1957); symbiotic star according to G. Wallerstein, *Obs* **101** (1981) 172.

WY Sge N

(N Sge 1783, N Sge No. 1)

Discovered by J. L. d'Agelet, who measured, in 1783, the position of a star missing from later catalogues (B. A. Gould, Catalogue of Stars Observed by d'Agelet, Washington Nat. Acad. Memoirs **1** (1966) 237).

Position:	19 30 29.70	+ 17 38 24.5	(POSS)
	19 30 29.68	+ 17 38 24.4	(2 recent observations)
	53.368	− 0.739	(G.C.)

Range: 6v – 19.5p LCT: ? t_3: ?

Finding chart: H. F. Weaver, *ApJ* **113** (1951) 320.
Duplicity: eclipsing binary with P = 0.1535 d (M. M. Shara, A. F. J. Moffat, J. T. McGraw, D. S. Dearborn, H. E. Bond, E. Kemper, R. Lamontagne, *ApJ* **292** (1984) 763).
Identification: from H. F. Weaver's finding chart.
Classification: nova; outburst characteristics hardly known; eclipses and bright and faint stages at minimum.

WZ Sge UGWZ

(N Sge 1913, 1946; N Sge No. 2, HV 3518)

Discovered by J. C. Mackie on Harvard plates (*HB* 691 (1919), *AN* **210** (1919) 79). Maxima occurred on 1913 November 22, 1946 June 28 (K. Himpel, *IAU Circ* 1054), and 1978 December 1 (J. T. McGraw, *IAU Circ* 3311).

Position:	20 05 20.58	+ 17 33 30.0	(POSS)
	20 05 20.64	+ 17 33 29.32	(A. Sh. Khatisov (1971))
	57.536	− 7.929	(G.C.)

Range: **7.0**p – 15.0p LCT: DN t_3: 30^d

Finding chart: M. Humason (1938); A. Sh. Khatisov (1971).

Light curve:	M. W. Mayall, *HB* 918 (1946) 3; W. Lohmann, G. R. Miczaika, *Heidelberg Ver* **14**, 9 (1946); M. Beyer, *AN* **280** (1951) 274; A. N. Eskioglu, *AAp* **26** (1963) 331; A. M. Heiser, G. W. Henry, *IBVS* 1559 (1979); E. Bohusz, A. Udalski, *IBVS* 1583 (1979); D. Targan, *IBVS* 1539 (1979); J. Bailey, *MN* **189** (1979) 41p; S. Ortolani, P. Rafanelli, L. Rosino, A. Vittone, *AsAp* **87** (1980) 31; J. A. Mattei, *JRAS Can* **74** (1980) 53.
Spectroscopy:	S. Ortolani, P. Rafanelli, L. Rosino, A. Vittone, *AsAp* **87** (1980) 31 – trac; D. Crampton, J. B. Hutchings, A. P. Cowley, *ApJ* **234** (1980) 182 – trac, rv.
	R. P. Kraft, *ApJ* **139** (1963) 457 – minimum spectrum, phot; G. Williams (1983) – minimum spectrum, trac.
UV observations:	A. C. Fabian, J. E. Pringle, D. J. Stickland, J. A. J. Whelan, *MN* **191** (1980) 457.
Duplicity:	R. P. Kraft, J. Mathews, J. L. Greenstein, *ApJ* **136** (1962) 312; W. Krzeminski, *PASP* **74** (1962) 66; W. Krzeminski, R. P. Kraft, *ApJ* **140** (1964) 921; W. Krzeminski, J. Smak, *AA* **21** (1971) 133.
Identification:	from Humason's finding chart.
Classification:	dwarf nova with long intervals between outbursts; previously classified as recurrent nova, the outburst behaviour is, however, typical for dwarf novae.

HS Sge **NA**

(N Sge 1977)

Discovered by J. G. Hosty, Huddersfield, England, on 1977 January 7, as a star of $7^{m}.2$ (*IAU Circ* 3025).

Position:	19 37 08.16	+ 18 00 57.45	(POSS)
	19 37 08.17	+ 18 00 58.0	(3 outburst observations)
	54.471	− 1.931	(G.C.)

Range: 7.0p – 20.5p LCT: ? t_3: 20^d

Light curve:	D. Boehme, *MVS* **8** (1977) 9, 33.
Spectroscopy:	F. C. Bruhweiler, H. A. Wotten, *BAAS* **9** (1977) 316 – descr; S. Wyckoff, H. Jenkner, *BAAS* **11** (1979) 462 – descr.
Identification:	from Harvard plate SH 5811, taken 1977 January 13, and the published positions.
Classification:	poorly known fast nova.

AT Sgr **NA:**

(N Sgr 1900, 268.1904 Sgr, HV 1149)

Discovered by H. S. Leavitt on Harvard plates (*HC* 91 (1904), *AN* **167** (1905) 165); nova nature proposed by H. H. Swope (*HB* 913 (1940)).

Position:	18 00 23.915	− 26 28 37.1	(SRC)
	18 00 23	− 26 28 41	(H. S. Leavitt, *HC* 91 (1904))
	4.127	− 2.172	(G.C.)

Range: 11.0p (10) – 19j LCT: ? t_3: 35^d

Finding chart:	H. H. Swope, *HA* **109** (1942) 1.
Light curve:	H. H. Swope, *HB* 913 (1940) 11.
Identification:	from Harvard plate A 4821, taken 1900 October 12.
Classification:	light curve form and amplitude suggest fast nova; no spectroscopic observations are available.

BS Sgr **NB**

(N Sgr 1917, Var. 4a, Zi 1390)

Discovered by R. T. A. Innes, Johannesburg Observatory. The star is first visible on a plate taken 1916 July 17, and reached maximum light on 1917 July 17 (*UOC* 37 (1917) 301). It was discovered independently by A. J. Cannon (*HC* 782 (1923)).

Position:	18 23 38.88	− 27 10 10.0	(SRC)
	18 23 42	− 27 08 40	(R. T. A. Innes, *UOC* 37 (1917) 301)
	5.997	− 7.062	(G.C.)

Range: **9.2**p – 17j LCT: ? t_3: 700^d

Light curve:	A. J. Cannon, *HB* 782 (1923); J. Dishong, D. Hoffleit, *AJ* **60** (1955) 259.
Spectroscopy:	H. W. Duerbeck, W. C. Seitter, *ApSS* **131** (1987) 467 – minimum spectrum, descr.
Identification:	from Harvard plate MF 2678, taken 1918 September 29, confirmed by spectroscopic observation at minimum. The nova is the brighter component of a close visual pair.
Classification:	poorly observed very slow nova.

FL Sgr

NA:

(N Sgr 1924, HV 4003)

Discovered by M. A. Gill on Harvard plates. The object is visible between 1924 May 13 and June 22 (*HB* 847 (1927)).

Position:			
(1)	17 57 10.74	– 34 36 11.4	(SRC)
(2)	17 57 10.03	– 34 36 12.3	(SRC, marked on chart)
(3)	17 57 10.19	– 34 36 04.1	(SRC)
	17 57 10	– 34 36 02	(M. A. Gill, *HB* 847 (1927))
	356.707	– 5.596	(G.C.)

Range: 8.3p (8) – 20j? LCT: ? t_3: 32^d

Light curve: M. A. Gill, *HB* 847 (1927); G. Cecchini, L. Gratton (1941) 127.

Identification: from Harvard plate MF 8724, taken 1924 July 25; the image quality does not allow a unambiguous identification, the three most likely candidates are given.

Classification: no spectroscopic observations are available; amplitude and light curve form suggest fast nova

FM Sgr

NA:

(N Sgr 1926, HV 3994)

Discovered by A. J. Cannon on Harvard plates. The object erupted between 1926 July 16 and 30, and remained brighter than 13^m until 1926 September 9. Maximum light is not covered by observations (*HB* 843 (1927)).

Position:			
(1)	18 14 15.16	– 23 39 36.7	(SRC)
(2)	18 14 15.365	– 23 39 35.4	(SRC)
	18 14 15	– 23 39 22	(A. J. Cannon, *HB* 843 (1927))
	8.117	– 3.547	(G.C.)

Range: 8.6p (8) – 20.5j? LCT: A or B t_3: 30^d

Light curve: A. J. Cannon, *HB* 843 (1927); C. Hoffmeister, *AN* **230** (1927) 183; J. Dishong, D. Hoffleit, *AJ* **60** (1955) 259; G. Cecchini, L. Gratton (1941) 135.

Identification: from Harvard plate A 14029, taken 1926 September 6; identification not unambiguous; the two best candidates are listed.

Classification: amplitude and light curve form suggest fast nova; no spectroscopic observations are available.

GR Sgr **N**

(N Sgr 1924, HV 4012)

Discovered by I. E. Woods on Harvard plates. The first plate, showing the nova at $11^{m}.4$, was taken 1924 April 30, obviously some months after maximum light (*HB* 851 (1927)).

Position:	18 19 52.88	– 25 36 20.6	(SRC)
	18 19 54	– 25 36 16	(I. E. Woods, *HB* 851 (1927))
	7.000	– 5.592	(G.C.)

Range: 11.4p (7.5) – 16.5 LCT: ? t_3: ?

Light curve: I. E. Woods, *HB* 851 (1927); J. Dishong, D. Hoffleit, *AJ* **60** (1955) 259.

Spectroscopy: H. W. Duerbeck, W. C. Seitter, *ApSS* **131** (1987) 467 – minimum spectrum; descr.

Identification: from Harvard plate MF 8762. The exnova is the brightest component of a blend of three stars.

Classification: the light curve is that of a slowly declining nova, maximum is not covered by observations.

HS Sgr **NB:**

(N Sgr 1900, HV 4011)

Discovered by I. E. Woods on Harvard plates taken in 1900 and 1901 (*HB* 851 (1927)).

Position:	18 25 03.50	– 21 36 20.6	(SRC)
	18 25 04	– 21 36 29	(I. E. Woods, *HB* 851 (1927))
	11.110	– 4.783	(G.C.)

Range: 11.5p (10.0) – 17: LCT: ? t_3: ?

Light curve: I. E. Woods, *HB* 851 (1927); G. Cecchini, L. Gratton (1941) 61.

Identification: from Harvard plate B 25879, taken 1900 September 1. The exnova has a faint SE companion, d = 3″.

Classification: the light curve is that of a very slow nova. No spectroscopic observations were obtained.

KY Sgr **NA:**

(N Sgr 1926, HV 4480)

Discovered by I. E. Woods on Harvard plates. Maximum light occurred 1926 June 11 or a few days earlier (*HB* 861 (1928) 5).

Position:	(1) 17 58 14.05	− 26 24 38.7	(SRC)
	(2) 17 58 13.89	− 26 24 38.6	(SRC)
	(3) 17 58 14.16	− 26 24 43.1	(SRC, marked on chart)
	17 58 14	− 26 24 14.6	(I. E. Woods, *HB* 861 (1928) 5)
	3.943	− 1.719	(G.C.)

Range: 10.6p (8?) − 20j LCT: B? t_3: 60^d

Finding chart: H. H. Swope, *HA* **109** (1942) 1.
Light curve: I. E. Woods, *HB* 861 (1928) 5, H. H. Swope, *HB* 913 (1940) 11; G. Cecchini, L. Gratton (1941) 134.
Identification: from Harvard plate MF 10366, taken 1926 June 14/15; the image quality does not allow an unambiguous identification; the three best candidates are listed.
Classification: amplitude and light curve form suggest moderately fast nova. No spectroscopic observations were obtained.

LQ Sgr N:

(N Sgr 1897, HV 4475)

Discovered by I. E. Woods, in 1928, on Harvard plates taken in 1897. Maximum light occurred 1897 September 21 (*HB* 859 (1928)).

Position:	18 25 19.93	− 27 57 17.1	(SRC)
	18 25 19	− 27 57 10	(I. E. Woods, *HB* 859 (1928))
	5.464	− 7.752	(G.C.)

Range: 13.0p − 21j LCT: ? t_3: ?

Light curve: I. E. Woods, *HB* 859 (1928); J. Dishong, D. Hoffleit, *AJ* **60** (1955) 259.
Identification: from Harvard plate A 2799, taken 1897 October 5.
Classification: amplitude and light curve form suggest nova. No spectroscopic observations were obtained.

V363 Sgr NA

(N Sgr 1927)

Discovered by F. Becker, La Paz, on an objective prism plate taken 1927 September 30. Maximum light occurred 1927 August 3 (*ZsAp* **1** (1930) 66).

Position:	(1) 19 08 05.81	− 29 55 01.45	(SRC)
	(2) 19 08 06.22	− 29 54 58.8	(SRC)
	19 08 11	− 29 57	(F. Becker, *ZsAp* **1** (1930) 66)
	7.629	− 17.072	(G.C.)

Range: 8.8p (8.7) – 20j LCT: A or Ba t_3: 64^d

Light curve: M. L. Walton, *HB* 878 (1930) 8; G. Cecchini, L. Gratton (1941) 144.
Spectroscopy: F. Becker, *ZsAp* **1** (1930) 66.
Identification: from Harvard plate MF 11393, taken 1928 February 25/26. Identification not unambiguous, the two most likely candidates are listed above.
Classification: moderately fast nova with smooth light curve, which is too scarcely covered by observations to reveal details, followed until late decline.

V441 Sgr NA

(N Sgr 1930, HV 5486, HD 315574)

Discovered by D. Hoffleit on Harvard plates. Maximum occurred 1930 September 12 (*HB* 890 (1932) 13).

Position:	18 19 02.68	− 25 30 23.6	(SRC)
	18 19 01	− 25 30 14	(D. Hoffleit, *HB* 890 (1932) 13)
	7.000	− 5.379	(G.C.)

Range: 8.7p (8.0) – ? LCT: Ba t_3: 53^d

Finding chart: A. J. Cannon, M. W. Mayall, *HA 112* (149) 189.
Light curve: D. Hoffleit, *HB* 890 (1932) 13; J. Dishong, D. Hoffleit, *AJ* **60** (1955) 259; G. Cecchini, L. Gratton (1941) 146.
Spectroscopy: D. Hoffleit, *HB* 890 (1932) 13.
Identification: from Harvard plate MF 14837, taken 1930 September 12. The image quality is too poor for identification of the exnova in the crowded field. The coordinates refer to one of a group of 7 stars which lie in the error-box.
Classification: moderately fast nova.

V522 Sgr N: or UG

(N Sgr 1931)

Discoverted by J. G. Ferwerda on plates taken by H. van Gent. Maximum occurred 1931 August 16/17 (*BAN* **7** (1935) 273)

Position: 18 44 55.60 − 25 25 44.3 (SRC)
18 44 55.7 − 25 25 34 (J. G. Ferwerda, *BAN* **7** (1935) 273)
9.175 − 10.565 (G.C.)

Range: 12.9p (12.8) − 17p LCT: ? t_3: ?

Finding chart: J. G. Ferwerda, *BAN* **7** (1935) 273; Yu. N. Efremov (1961).
Light curve: J. G. Ferwerda, *BAN* **7** (1935) 273.
Identification: from Ferwerda's and Efremov's finding charts; blue object
Classification: light curve obtained in four consecutive nights is available. Dwarf nova type is possible in view of the small amplitude and lack of spectroscopic verification.

V630 Sgr NA

(N Sgr 1936, N Sgr No. 7, 619.1936 Sgr, HD 321353)

Discovered by S. Okabayashi, Kobe, Japan, 1936 October 3, as a star of $4^{m}.5$ (*IAU Circ 622, AN* **261** (1926) 65).

Position: 18 05 28.94 − 34 20 52.65 (SRC)
18 05 29.10 − 34 20 52.05 (2 outburst observations)
357.768 − 6.967 (G.C.)

Range: **4.5**v − 19j LCT: Ao t_3: 11^d

Finding chart: A. J. Cannon, M. W. Mayall, *HA* **112** (1949) 204; J. Ponsen, *Leiden Ann* **20**, 17 (1957).
Light curve: E. Hertzsprung, *IAU Circ* 1194 (1948); P. P. Parenago, *PZv* **7** (1949) 109; S. Gaposchkin, *AJ* **60** (1955) 454; J. Ponsen, *Leiden Ann* **20**, (1957) 17, G. Cecchini, L. Gratton (1941) 166; C. Payne-Gaposchkin (1957) 152.
Spectroscopy: A. H. Joy, W. S. Adams, T. Dunham, Jr., *PASP* **48** (1936) 328 − descr, rv; A. S. Wyse, *PASP* **49** (1937) 290 − descr; C. Payne-Gaposchkin (1957) 152.
Identification: from Harvard plate MF 22541, taken 1936 October 9/10, and from the published accurate positions, spectroscopically confirmed by S. Dieters (private communication).
Classification: well-observed very fast nova.

V726 Sgr **NA**

(N Sgr 1936, 171.1937 Sgr, 18.1938 Sgr, HV 9380, HD 315532)

Discovered by M. W. Mayall, and independently by W. Luyten. It is invisible until 1936 April 30, maximum light was reached 1936 May 13 (*HAC* 439 (1937), *HB* 908 (1938), *AN* **264** (1937) 63).

Position:	18 16 26.22	– 26 54 38.2	(SRC)
	18 16 27	– 26 54 49	(M. W. Mayall, *HB* 908 (1938), W. Luyten, *AN* **264** (1937) 63)
	5.479	– 5.519	(G.C.)

Range: 10.8p (10.5) – 19j? LCT: Cb? t_3: 95^d

Finding chart: A. J. Cannon, M. W. Mayall, *HA* **112** (1949) 189.
Light curve: M. W. Mayall, *HB* 907 (1938) 28; W. Luyten, *AN* **264** (1937) 63; J. Dishong, D. Hoffleit, *AJ* **60** (1955) 259.
Spectroscopy: M. W. Mayall, *HB* 908 (1938).
Identification: from Harvard plate B 61025, taken 1936 June 25/26; identification is not unambiguous; the best candidate of a group of about 7 stars is listed above.
Classification: moderately fast nova with strong [Ne III] lines.

V732 Sgr **NA**

(N Sgr 1936, 225.1937 Sgr, HD 316633)

Discovered by C. G. Burwell on a Mt. Wilson objective prism plate taken 1936 June 10 (*PASP* **49** (1937) 342).

Position:	17 52 59.18	– 27 21 53.2	(SRC; approximate position)
	17 52 58	– 27 22 06	(L. Plaut, *Leiden Ann* **20** (1948) 1)
	2.526	– 1.186	(G.C.)

Range: 6.5p – ? LCT: Ca t_3: 74^d

Finding chart: A. J. Cannon, M. W. Mayall, *HA* **112** (1949) 192; L. Plaut, *Leiden Ann* **20** (1948) 1.
Light curve: R. B. Jones, *HB* 907 (1938) 29; K. Himpel, *AN* **272** (1941) 80; H. H. Swope, *HB* 913 (1940) 11; P. Ahnert, *MVS* 37 (1943); L. Plaut, *Leiden Ann* **20** (1948) 1; G. Cecchini, L. Gratton (1941) 155; C. Payne-Gaposchkin (1957) 13, 14.
Spectroscopy: C. G. Burwell, *PASP* **49** (1937) 342.

Identification:	from Harvard plate B 61199, taken 1936 August 11/12; identification and derivation of the light curve are complicated because of the near coincidence of the nova with CoD −27° 12120. The image of the Harvard plate seems to coincide with a 16^m star whose coordinates are given above; its spectroscopic appearance is unlike that of an exnova. A study of the stars in the immediate surroundings would be useful.
Classification:	well-observed moderately fast nova of DQ Her type with light echo; in obscured region of the Galaxy.

V737 Sgr N:

(V908 Sgr, N Sgr 1933, 201.1937 Sgr)

Discovered by W. O'Leary, Riverview College Observatory, Australia. Maximum light occurred probably around 1933 June 30 (*AN* **264** (1937) 141).

Position:	18 03 58.335	−28 45 16.6	(SRC)
	18 03 58	−28 45 36	(W. O'Leary, *AN* **264** (1937) 141)
	2.531	− 3.982	(G.C.)

Range: 10.3p (7 ... 10) − 19j? LCT: Ca? t_3: $>70^d$

Light curve:	W. O'Leary, *Riv Publ* **1** (1937) 56.
Identification:	from Harvard plate RB 4299, taken 1933 June 28/29, small scale, tentative identification. The exnova appears oval and is obviously a blend of several stars.
Classification:	the light curve shows a plateau of at least 70 days; the rise and decline were not observed. No spectroscopic observations are available. Possible nova.

V787 Sgr NA:

(N Sgr 1937, HV 10322, HD 316917)

Discovered by H. H. Swope on Harvard plates. Maximum occurred near the end of 1937 May (*HB* 913 (1940)).

Position:	(1)	17 56 49.02	−30 30 23.1	(SRC, star 20^m)
	(2)	17 56 49.01	−30 30 26.1	(SRC, star 21^m)
	(3)	17 56 49.18	−30 30 21.0	(SRC, star 21^m)
		17 56 47	−30 30 25	(L. Plaut, *Leiden Ann* **20** (1948) 1)
		0.235	− 3.494	(G.C.)

Range: 9.8p – 21j? LCT: Ba t_3: 74^d

Finding chart: L. Plaut, *Leiden Ann* **20**, 1 (1949); A. J. Cannon, M. W. Mayall, *HA* **112** (1949) 192.
Light curve: H. H. Swope, *HB* 913 (1940) 11; L. Plaut, *Leiden Ann* **20** (1948) 1; D. J. K. O'Connell, *Acta Pontif Acad Sci* **16** (1954) 49 = *Riv Repr* 8; C. Payne-Gaposchkin (1957) 11.
Identification: from Harvard plate B 61984, taken 1937 May 19/20. Identification is not unambiguous; three candidate stars are listed.
Classification: amplitude and light curve suggest moderately fast nova. No spectroscopic observations are available.

V908 Sgr

see V737 Sgr.

V909 Sgr **NA**

(N Sgr 1941)

Discovered by M. W. Mayall on Harvard objective prism plates taken 1941 July 18 and 19. Maximum light was reached 1941 June 26 ((*PA* **51** (1943) 284).

Position:			
(1)	18 22 32.26	– 35 03 13.0	(SRC)
(2)	18 22 31.92	– 35 03 12.3	(SRC)
	18 22 32.7	– 35 03 17	(M. W. Mayall, *HB* 918 (1946) 1)
	358.767	– 10.404	(G.C.)

Range: **6.8**p – 20j LCT: A t_3: 7^d

Light curve: M. W. Mayall, *HB* 918 (1946) 1; L. Campbell, *Harv Repr* 259 (1943) 15; C. Payne-Gaposchkin (1957) 9.
Spectroscopy: M. W. Mayall, *HB* 918 (1946) 1.
Identification: From Harvard plate MF 29346, taken 1941 July 1/2; identification ambiguous; the positions of the components the close visual binary are listed above.
Classification: very fast nova.

V927 Sgr **NA**

(N Sgr 1944)

Discovered by M. W. Mayall on Harvard objective prism plates. Maximum light was reached on 1944 April 16 (*PA* **55** (1947) 109).

Position: 18 04 25.45 − 33 21 43.9 (SRC)
18 04 24 − 33 21 48 (M. W. Mayall, *PA* **55** (1947) 109)
358.532 − 6.302 (G.C.)

Range: 8.0p (7.3) – 20j? LCT: A? t_3: $\leq 15^d$

Light curve: L. Campbell, *PA* **55** (1947) 109.
Identification: from Harvard plate MF 32371, taken 1944 April 21/22; identification not unambiguous due to crowded field and poor image quality of plate. The nova could be one of about 7 stars in the $\pm 5''$ error box.
Classification: poorly known very fast nova.

V928 Sgr NB

(N Sgr 1947)

Discovered by C. G. Burwell on a Mt. Wilson objective prism plate taken 1947 May 16 (*IAU Circ* 1092)

Position: (1) 18 15 49.83 − 28 07 14.9 (SRC, star 21^m)
(2) 18 15 50.20 − 28 07 17.0 (SRC, star 21^m)
18 15 52.2 − 28 07 22 (G. Merton, *IAU Circ* 1094 (1947))
4.342 − 5.971 (G.C.)

Range: 8.9p (8.5) – 20.5j LCT: B? t_3: 150^d

Light curve: C. Bertaud, M. Herman, *JO* **30** (1947) 5; J. Warren (unpublished).
Spectroscopy: P. W. Merrill, C. G. Burwell, W. C. Miller, *PASP* **59** (1947) 194 – trac, descr.
Identification: from Harvard plate MF 36384, taken 1947 June 11/12, no unambiguous identification possible.
Classification: slow nova; radial velocity + 190 km s^{-1} from bright lines.

V939 Sgr M

(N Sgr 1914, Innes 25, Zi 1428)

Discovered by R. T. A. Innes, Johannesburg Observatory. The object is first seen on a plate of 1913 August, it reached maximum light in 1914, and faded in 1915 (*UOC* 20 (1914) 152).

Position: 18 31 15 − 26 56 30 (R. T. A. Innes, *UOC* 20 (1914) 152)
6.973 − 8.469 (G.C.)

Range: 14.2p – ? LCT: M? t_3: –

Finding chart: D. Hoffleit, *AJ* **66** (1016) 188.
Light curve: R. T. A. Innes, *UOC* **37** (1917) 302.
Identification: from Hoffleit's finding chart and a comparison of recent sky atlas plates.
Classification: D. B. McLaughlin (*AJ* **51** (1945) 139) classified the star as nova of RT Ser type, D. Hoffleit (*AJ* **66** (1961) 188) as Mira variable with $P = 336^d$; the latter classification is supported by spectroscopy (H. W. Duerbeck, in preparation).

V941 Sgr **M**

(N Sgr 1912, Innes 27, Zi 1430)

Discovered by R. T. A. Innes, Johannesburg Observatory. The object increased in brightness between 1910 July 29 and September 1, maximum light occurred 1912 June 21, followed by a decline with superimposed brightness variations. After 1916 July 19, the star disappeared (*UOC* 20 (1914) 152).

Position: 18 31 32.12 − 29 37 13.5 (SRC)
18 31 32 − 29 36 49 (R. T. A. Innes, *UOC* 20 (1914) 152)
4.570 − 9.712 (G.C.)

Range: 11.0p – ? LCT: M? t_3: –

Finding chart: S. W. McCluskey, R. Mehlhorn, *AJ* **68** (1963) 319.
Light curve: R. T. A. Innes, *UOC* 37 (1917) 302.
Identification: from McCluskey and Mehlhorn's finding chart.
Classification: D. B. McLaughlin (*AJ* **51** (1945) 139) classified the object as a nova of RT Ser Type. S. W. McCluskey and R. Mehlhorn (*AJ* **68** (1963) 319) find M type spectrum and suspect Mira variability. A recent spectroscopic observation yields an M type spectrum without indications of symbiotic character (H. W. Duerbeck, in preparation).

V949 Sgr **N:**

(N Sgr 1914, Innes 46, Zi 1468)

Discovered by R. T. A. Innes, Johannesburg Observatory. The object is visible on three plates taken between 1914 July 16 and July 25 (*UOC* 20 (1914) 152).

Position: 18 37 55 − 28 12 20 (R. T. A. Innes, *UOC* 37 (1917) 202)
6.482 − 10.351 (G.C.)

Range: 15.7p – ? LCT: ? t_3: ?

Light curve: R. T. A. Innes, *UOC* 37 (1917) 202.
Identification: not made because of lack of precise position and finding chart. Field map in Appendix.
Classification: probably a faint nova. No spectroscopic information is available.

V990 Sgr NA:

(Plaut No. 57, N Sgr 1936)

Discovered by L. Plaut, Leiden, on plates taken by H. van Gent. Maximum light occurred on 1936 September 12 (*Leiden Ann* **20** (1948) 30).

Position: 17 54 09 − 28 18 48 (L. Plaut, *Leiden Ann* **20** (1948) 30)
1.842 − 1.890 (G.C.)

Range: **11.1**p – ? LCT: A? t_3: 24^d

Finding chart: L. Plaut, *Leiden Ann* **20** (1948) 30.
Light curve: L. Plaut, *Leiden Ann* **20** (1948) 30.
Identification: from Harvard plate B 61379, taken 1936 September 20/21.
Classification: light curve of a fast nova. No spectroscopic information is available.

V999 Sgr NB

(N Sgr 1910, N Sgr No. 2, 96.1910 Sgr, HV 3304, HD 163982)

Discovered by W. Fleming on Harvard plates. The nova is visible between 1910 March 21 and June 10 (*AN* **186** (1910) 125).

Position: 17 56 57.05 − 27 33 07.5 (SRC)
17 56 57.08 − 27 33 07.15 (3 outburst observations)
2.811 − 2.042 (G.C.)

Range: **8.0**p – 17.35B LCT: D t_3: 160^d

Finding chart: A. J. Cannon, M. W. Mayall, *HA* **112** (1949) 192.
Light curve: A. D. Walker, *HA* **84** (1923) 189; H. H. Swope, *HB* 913 (1940) 11; G. Cecchini, L. Gratton (1941) 88; C. Payne-Gaposchkin (1957) 16, 201.

Spectroscopy: W. H. Wright, *Lick Bull* **6** (1910) 65 – descr; A. J. Cannon, *HA* **76** (1916) 19 – descr; H. W. Duerbeck, W. C. Seitter, *ApSS* **131** (1987) 467 – minimum spectrum, descr.
Identification: from Harvard plate A 10392, taken 1911 April 29.
Classification: slow nova.

V1012 Sgr NA:

(N Sgr 1914, N Sgr No. 7, HV 3531)

Discovered by I. E. Woods on Harvard plates. Maximum occurred on 1914 August 12 (*HB* 733 (1920), *AN* **213** (1920) 47).

Position:			
(1)	18 02 59.09	– 31 44 46.05	(SRC)
(2)	18 02 58.98	– 31 44 48.45	(SRC)
	18 02 59	– 31 44 48	(S. I. Bailey, *AN* **213** (1920) 47)
	359.806	– 5.254	(G.C.)

Range: **8.0**p – 20j LCT: A or Ba t_3: 32^d

Light curve: S. I. Bailey, *HB* 733 (1920).
Identification: from Harvard plate B 44946, taken 1914 August 12. The nova image coincides with two stars of 20^m and 21^m.
Classification: fragmentary light curve; amplitude and light curve form suggest fast nova. No spectroscopic information is available.

V1014 Sgr NA:

(N Sgr 1901, N Sgr No. 4, 22.1911 Sgr, HV 3312)

Discovered by A. J. Cannon on Harvard plates. The object is visible between 1901 May 22 and July 9 (*HC* 164 (1911), *AN* **188** (1911) 259).

Position:			
(1)	18 03 37.28	– 27 26 36.7	(SRC)
(2)	18 03 37.45	– 27 26 39.9	(SRC)
	18 03 35.6	– 27 26 21	(E. C. Pickering, *HC* 164 (1911))
	3.639	– 3.274	(G.C.)

Range: **10.9**p – 20j LCT: D or Ca t_3: $> 50^d$

Light curve: A. D. Walker, *HA* **84** (1923) 197; G. Cecchini, L. Gratton (1941) 71; C. Payne-Gaposchkin (1957) 201.

Identification:	from Harvard plate B 27750, taken 1901 June 25; two candidates are listed above.
Classification:	fairly slow nova; fragmentary light curve. No spectroscopic information is available.

V1015 Sgr NA:

(N Sgr 1905, N Sgr No. 6, HV 3529)

Discovered by I. E. Woods on Harvard plates. Maximum occurred on 1905 July 27 (*HB* 714 (1920), *AN* **213** (1921) 47)

Position:			
	18 05 45.78	− 32 29 04.1	(SRC; best candidate, 21^{m})
	18 05 47	− 32 28 48	(I. E. Woods, *HB* 714 (1920))
	359.442	− 6.126	(G.C.)

Range: 7.1p (6.5) – 21j? LCT: A or B t_3: 34^{d}

Light curve:	S. I. Bailey, *HB* 714 (1920).
Identification:	from Harvard plates AM 3802, taken 1905 August 18, and B 36792, taken 1905 August 22. Crowded field; the best candidate is listed.
Classification:	fragmentary light curve; amplitude suggests nova. No spectroscopic observations are available.

V1016 Sgr NB:

(N Sgr 1899, N Sgr No. 3, 132.1901 Sgr, HV 3306)

Discovered by A. J. Cannon on Harvard plates. The nova is $8\overset{m}{.}5$ between 1899 August 10 and 23 (*HC* 163 (1911), *AN* **186** (1910) 319).

Position:			
	18 16 52.53	− 25 12 35.1	(SRC)
	18 16 52	− 25 12 23	(A. J. Cannon, *HC* 163 (1911))
	7.032	− 4.806	(G.C.)

Range: 8.5p (8.4) –17p LCT: A? t_3: 140^{d}

Light curve:	A. J. Cannon, *AN* **188** (1911) 77; A. D. Walker, *HA* **84** (1923) 189; G. Cecchini, L. Gratton (1941) 60, 61; J. Dishong, D. Hoffleit, *AJ* **60** (1955) 259.
Identification:	from Harvard plate B 24376, taken 1899 October 13. The nova has two fainter companions, 21^{m} and 23^{m}.
Classification:	light curve of a slow nova; no spectroscopic observations are available.

V1017 Sgr **NR**

(N Sgr 1919, N Sgr No. 5, HV 3519)

Discovered by I. E. Woods on Harvard plates. The nova is 15^m at minimum; it brightened to 7^m on 1919 March 11; brightenings to 11^m occurred in 1901 and 1973 (I. E. Woods, *HB* 693 (1919), *AN* **210** (1919) 79).

Position:	18 28 53.30	– 29 25 25.7	(SRC)
	18 28 53.44	– 29 25 25.7	(G. Williams (1983))
	4.490	– 9.109	(G.C.)

Range: **7.2**p – 14.7B LCT: Ba? pec? t_3: 130^d

Finding chart: M. Humason (1938); G. Williams (1983); N. V. Vidal, A. W. Rodgers, *PASP* **86** (1974) 26.

Light curve: D. B. McLaughlin, *PASP* **58** (1946) 46; J. A. Mattei, *JRAS Can* **68** (1974) 221; S. J. Kenyon (1986) 239.

Spectroscopy: N.V. Vidal, A. W. Rodgers, *PASP* **86** (1974) 26 – phot, ident (outburst); H. W. Duerbeck, W. C. Seitter, *ApSS* **131** (1987) 467 – minimum spectrum (descr), two component (symbiotic) spectrum; R. P. Kraft, *ApJ* **139** (1964) 457 – minimum spectrum, phot; G. Williams (1983) – minimum spectrum, trac.

Classification: recurrent nova with slowly developing outburst of different amplitude; possibly symbiotic.

V1059 Sgr **NA**

(N Sgr 1898, N Sgr No. 1, HV 129, IC 4816, HN 81, HD 176654)

Discovered by W. Fleming on Harvard plates. The nova was invisible until 1897 October 23, and appears first on a plate taken 1898 March 8 (*HC* 42, *AN* **149** (1899) 13).

Position:	18 59 01.62	– 13 14 03.7	(SRC)
	18 59 01.72	– 13 14 03.1	(3 outburst observations)
	22.305	– 8.230	(G.C.)

Range: 4.9p (2.0) – 18.1B LCT: A? t_3: ?

Finding chart: M. Humason (1938); A. Sh. Khatisov (1971).

Light curve: A. D. Walker, *HA* **84** (1923) 189; G. Cecchini, L. Gratton (1941) 56, 57.

Spectroscopy: E. C. Pickering, *HC* 42 (1899) – descr; W. P. Fleming, *HA* **56** (1912)

	165 – descr; A. J. Cannon, *HA* **76** (1916) 19 – descr; H. W. Duerbeck, W. C. Seitter, *ApSS* **131** (1987) 467 – minimum spectrum, descr.
Identification:	from Harvard plates A 4086, taken 1899 October 25, and A 4091, taken 1899 October 26. The exnova is the brightest of a blend of three stars; a slightly fainter component is $3\overset{m}{.}5$ SW, a much fainter component is $6\overset{m}{.}6$ SE.
Classification:	poorly observed fast nova.

V1148 Sgr N

(N Sgr 1948)

Discovered by M. W. Mayall on Harvard objective prism plates taken 1943 August 3. The nova showed a K type spectrum at maximum and is close to the globular cluster NGC 6553 (*AJ* **54** (1949) 191).

Position:			
	18 05 59.5	– 25 59 40	(SRC; brightest candidate, 14^m)
	18 06 00	– 26 00 05	(M. W. Mayall, *AJ* **54** (1949) 191)
	5.164	– 3.028	(G.C.)

Range: 8.0 p – ? LCT: ? t_3: ?

Spectroscopy:	M. W. Mayall, *AJ* **54** (1949) 191 – descr.
Identification:	the nova is recorded on Harvard plates MF 31866, 31871 and 31911 which are not available at Harvard. The plates RB 12324, taken 1943 August 2/3, and RB 12347, taken 1943 August 24/25, which have a smaller scale, were used. The identification is ambiguous; the brightest candidate is listed above, which is very probably not the exnova.
Classification:	nova; spectroscopic observations yielded a K type spectrum at maximum, which developed into a typical bright line spectrum during later phases.

V1149 Sgr NB

(N Sgr 1945)

Discovered by M. W. Mayall on Harvard objective prism plates. The nova is first seen on 1945 February 16 as a 9^m star. According to J. Warren, it is visible between 1945 May 16 to 1948 May 30 (*AJ* **54** (1949) 191; *AAVSO Abstr* October 1965, 8).

Position:			
	18 15 20.81	– 28 18 31.4	(SRC)
	18 15 20	– 28 18 36	(M. W. Mayall, *AJ* **54** (1949) 191)
	4.124	– 5.962	(G.C.)

Range: 7.4p – 21j LCT: Ba? t_3: <210^d

Light curve: L. Campbell, *PA* **55** (1949) 392 = *Harv Repr* 300 (1949) 22; J. Warren, *AAVSO Abstr* October 1965, 8.

Identification: from Harvard plates MF 34497, taken 1945 August 1/2, and MF 34530, taken 1945 August 6/7; tentative identification; two stars of similar magnitudes in the vicinity are also possible candidates.

Classification: slow nova.

V1150 Sgr — NB

(N Sgr 1946)

Discovered by M. W. Mayall on Harvard objective prism plates. The initial rise was not covered, observations following 1946 May 28 show a plateau (*AJ* **54** (1949) 191).

Position:			
(1)	18 15 51.15	– 24 06 46.4	(SRC)
(2)	18 15 51.53	– 24 06 45.1	(SRC)
	18 15 51	– 24 07 00	(M. W. Mayall, *AJ* **54** (1949) 191)
	7.892	– 4.084	(G.C.)

Range: 13.3p (12) – [22j? LCT: ? t_3: <600^d

Light curve: J. Warren, *AAVSO Abstr* October 1965, 8.

Identification: from Harvard plates MF 35502 and MF 35504, taken 1946 July 1/2. The nova coincides probably with the empty field between the two stars listed above.

Classification: very slow nova; the observed maximum lasts for 90^d, it is similar to that of DO Aql.

V1151 Sgr — NB

(N Sgr 1947)

Discovered by M. W. Mayall on Harvard objective prism plates. Maximum occurred on 1947 April 17 (*AJ* **54** (1949) 191).

Position:		
18 22 25.60	– 20 13 43.7	(SRC)
18 22 22	– 20 13 35	(M. W. Mayall, *AJ* **54** (1949) 191)
12.045	– 3.598	(G.C.)

Range: 11.1p (10.5) – 20j LCT: B or Cb t_3: 135^d

Light curve: J. Warren, unpublished.
Identification: from Harvard plates B 72630 and MF 36228, taken 1947 May 14/15, and MF 36256, taken 1947 May 18/19.
Classification: slow nova; maximum poorly covered.

V1172 Sgr N

(N Sgr 1951)

Discovered by G. Haro, Tonantzintla Observatory, 1951 March 7, as a 9^m star (*IAU Circ* 1306, *HAC* 1118).

Position:	17 47 24.69	−20 39 41.8	(SRC)
	17 47 24.6	−20 39 44	(G. Haro, *IAU Circ* 1306 (1951))
	7.639	+ 3.335	(G.C.)

Range: 9.0p – 18j LCT: ? t_3: ?

Spectroscopy: G. Haro, *HAC* 1118 (1951), *HAC* 1119 (1951) – descr.
Identification: from Haro's precise position; independent confirmation is necessary.
Classification: poorly known nova.

V1174 Sgr N

(N Sgr 1952 No. 2)

Discovered by G. Haro, Tonantzintla Observatory, 1952 March 29; the outburst occurred 1952 February 21 or shortly before (*IAU Circ* 1353, *HAC* 1172).

Position:	17 58 27	−28 44 26	(G. Haro, *IAU Circ* 1353 (1952))
	1.946	− 2.922	(G.C.)

Range: 12.0p – ? LCT: ? t_3: ?

Identification: no finding chart or accurate position is available. Identification not possible. Field map in Appendix.
Classification: poorly known nova.

V1175 Sgr N

(N Sgr 1952 No. 1)

Discovered by G. Haro, Tonantzintla Observatory, 1952 February 21 (*IAU Circ* 1347, *HAC* 1166).

Position: 18 11 03   − 31 08 03   (G. Haro, *IAU Circ* 1347 (1952))
1.175   − 6.470   (G.C.)

Range: 7.0 − ?   LCT: ? t_3: ?

Light curve: D. Taboada, *Ton Bol* **5** (1952).
Spectroscopy: M. Feast, *MNASSA* **11** (1952) 51 − descr, iden, rv.
Identification: no finding chart or accurate position is available; identification not possible. Field map in Appendix.
Classification: poorly known nova.

V1274 Sgr **N**

(N Sgr 1954 No. 2)

Discovered by P. Wild, Mt. Palomar Observatory, 1954 August 30, as a star of $10^{m}.5$. First found on films taken 1954 August 2 and 3 as a 12^{m} star (*IAU Circ* 1471).

Position: 17 46 00   − 17 51   (P. Wild, *IAU Circ* 1471 (1954))
9.885   + 5.067   (G.C.)

Range: 10.5 − ?   LCT: ? t_3: ?

Spectroscopy: K. Yoss, *HAC* 1271 (1954) − descr.
Identification: no finding chart or accurate position is available; identification not possible. Field map in Appendix.
Classification: poorly known nova.

V1275 Sgr **NA**

(N Sgr 1954 No. 1, N Sco 1954)

Discovered by G. Haro and L. Herraro, Tonantzintla Observatory, 1954 July 4. The nova was fainter than 13^{m} on 1954 July 1 (*IAU Circ* 1459).

Position: 17 55 43.44   − 36 18 29.1   (SRC)
17 55 43.41   − 36 18 29.85   (I. Mitani, *IAU Circ* 1469 (1954))
355.069   − 6.182   (G.C.)

Range: **7.0**p − 18j?   LCT: ? t_3: $> 10^{d}$

Spectroscopy: M. W. Feast, *MN* **115** (1955) 461 − descr, ident, rv; H. A. Abt, *ApJ* **122** (1955) 199 − phot.
Identification: from I. Mitani's precise position; double star.
Classification: fast nova.

Spectroscopy: D. A. Allen (1984) – trac.
Identification: from Harvard plates B 71556, taken 1946 April 10/11, and B 71709, taken 1946 May 28/29.
Classification: symbiotic star.

V2572 Sgr NA:

(N Sgr 1969, BV 1262)

Discovered by F. M. Bateson and R. G. Welch on Bamberg plates taken in 1969 June–September. Maximum occurred near the beginning of July (*IBVS* 389 (1969)).

Position:			
	18 28 20.68	– 32 38 08.4	(SRC)
	18 28 20.58	– 32 38 10.1	(2 Bamberg sky survey plates, ± 5″)
	1.514	– 10.422	(G.C.)

Range: **6.5**p – 18j LCT: Ba t_3: 44^d

Finding chart: J. R. Sievers, *IBVS* 452 (1970); P. R. Knight, *IBVS* 694 (1972).
Light curve: F. M. Bateson, B. F. Marino, W. S. G. Walker, *IBVS* 475 (1970); P. R. Knight, *IBVS* 694 (1972); I. Radiman, B. Hidajat, *IBVS* 976 (1975).
Identification: from Bamberg plates NZ 1774, taken 1969 July 8, and NZ 1954, taken 1969 August 16. The exnova is a blend of three stars, the coordinates refer to the brightest, central star.
Classification: amplitude and light curve form suggest moderately fast nova. No spectroscopic observations are available.

V3645 Sgr NB

(N Sgr 1970, SVS 1728)

Discovered by V. Archipova and O. Dokuchaeva on an objective prism plate taken by R. Bartaya and T. Vashakidse, Abastumani Observatory, 1970 July 29. Maximum light occurred during the winter of 1969/1970 (*IBVS* 494 (1970)).

Position:			
	18 32 53.15	– 18 44 14.7	(SRC)
	18 32 53.39	– 18 44 25.7	(A. Sarajedini, *IBVS* 2587 (1984))
	14.515	– 5.093	(G.C.)

Range: 12.6 p (8?) – 18p LCT: D? t_3: 300^d?

Finding chart: V. P. Archipova, O. D. Dokuchaeva, T. G. Nikulina, *PZv* **18** (1971) 195.

Light curve: V. P. Archipova, O. D. Dokuchaeva, T. G. Nikulina, *PZv* **18** (1971) 195; A. Sarajedini, *IBVS* 2587 (1984).
Spectroscopy: V. P. Archipova, O. D. Dokuchaeva, T. G. Nikulina, *PZv* **18** (1971) 195 – trac.
Identification: from Maria Mitchell Observatory plate NA 4803, taken 1969 September 11; the center of the outburst image coincides with a group of 5 stars; the most likely candidate is listed above.
Classification: slow nova; maximum missed.

V3876 Sgr — M

(N Sgr 1978, IRC-20494)

Discovered by M. Honda, Japan, 1978 April 7, as a 10^m star (*IAU Circ* 3209).

Position:			
	18 30 14.68	– 20 08 07.1	(SRC)
	18 30 14.72	– 20 08 08.0	(2 outburst observations)
	12.980	– 5.180	(G.C.)

Range: 11.6p – [14.2p LCT: M t_3: –

Finding chart: D. Hoffleit, *IBVS* 729 (1972).
Spectroscopy: C. B. Stephenson, N. Sanduleak, *IBVS* 1213 (1976).
Identification: from published finding charts and precise positions.
Classification: Mira variable with P = 345 or 358^d (D. Hoffleit, *IBVS* 729 (1972).

V3888 Sgr — N

(N Sgr 1974)

Discovered by Y. Kuwano, Japan, 1976 October 6, as a 9^m star (*IAU Circ* 2707).

Position:			
	17 45 44.74	– 18 44 41.1	(SRC)
	17 45 44.08	– 18 44 41.3	(3 outburst observations)
	9.084	+ 4.659	(G.C.)

Range: 9.0v (6.5) – 16j LCT: ? t_3: ?

Finding chart: R. F. Schmidt, *IBVS* 946 (1974); A. Sh. Khatisov, *ATs* 852 (1975) 7; E. M. Leibowitz, S. Wyckoff, N. V. Vidal, *PASP* **88** (1976) 750.
Light curve: N. Vogt, H. M. Maitzen, *AsAp* **61** (1977) 601; E. M. Leibowitz, S. Wyckoff, N. V. Vidal, *PASP* **88** (1976) 750.

Spectroscopy: E. M. Leibowitz, S. Wyckoff, N. V. Vidal, *PASP* **88** (1976) 750 – ident, phot, rv.
Identification: from published precise positions; the exnova is slightly elongated (double star).
Classification: nova; discovered during Orion phase.

V3889 Sgr — NA:

(N Sgr 1975)

Discovered by Y. Kuwano, Japan, 1975 July 13 (*IAU Circ* 2805).

Position:			
	17 55 11.57	– 28 21 38.6	(SRC)
	17 55 11.65	– 28 21 36.8	(3 outburst observations)
	1.917	– 2.113	(G.C.)

Range: 8.4v – 21j LCT: A? t_3: 14^d

Finding chart: A. Sarajedini, *IBVS* 2587 (1984).
Light curve: A. Sarajedini, *IBVS* 2587 (1984).
Identification: from Maria Mitchell Observatory plate NA 5589, taken 1975 August 10, published accurate positions, and from Sky Atlas plate ESO QB 456, taken 1976 April 23 (during decline).
Classification: amplitude and light curve form suggest fast nova, whose maximum was barely missed. No spectroscopic observations are available.

V3890 Sgr — N or ZAND

(N Sgr 1962)

Discovered by H. Dinerstein on plates taken at the Maria Mitchell Observatory by D. Hoffleit. The maximum magnitude was $8^{m}_{.}4$, but maximum light occurred probably during the gap between 1962 May 10 and June 2 (H. Dinerstein, *IBVS* 845 (1973)).

Position:			
	18 27 39.66	– 24 03 05.8	(SRC)
	18 27 40.47	– 24 03 05.9	(G. Williams (1983))
	9.206	– 6.441	(G.C.)

Range: 8.4 p (4...8) – 15.4...17.0 LCT: ? t_3: ?

Finding chart: H. Dinerstein, *IBVS* 845 (1973); G. Williams (1983).
Light curve: H. Dinerstein, *IBVS* 845 (1973).
Spectroscopy: G. Williams (1983) – minimum spectrum, trac (symbiotic star?)

Identification: from published finding charts.
Classification: nova without spectroscopic confirmation at outburst; exnova shows dwarf nova-type variability.

V3964 Sgr NA

(N Sgr 1975 No. 2)

Discovered by I. Lundström and B. Stenholm on an objective prism plate taken with the Uppsala Southern Schmidt telescope, 1975 June 8 (*IAU Circ* 2997 (1977)).

Position:			
	17 46 47.77	– 17 22 44.3	(SRC)
	17 46 47.4	– 17 22 30	(I. Lundström, B. Stenholm, *IBVS* 1351 (1977))
	10.389	– 5.146	(G.C.)

Range: 9.4p (6) – [17p LCT: Ba? t_3: 32^d

Finding chart: I. Lundström, B. Stenholm, *IBVS* 1351 (1977); V. P. Goranskij, *ATs* 983 (1978) 1.
Light curve: V. P. Goranskij, *ATs* 983 (1978) 1.
Identification: from Goranskij's finding chart. Lundström and Stenholm's finding chart is incorrect.
Classification: fast nova.

V4021 Sgr NA

(N Sgr 1977)

Discovered by Y. Kuwano, Japan, 1977 March 27 (*IAU Circ* 3055).

Position:			
	18 35 12.03	– 23 25 26.95	(SRC)
	18 35 11.82	– 23 25 26.5	(3 outburst observations)
	10.555	– 7.696	(G.C.)

Range: 8.8v (7.5) – 18j LCT: B? t_3: 70^d

Finding chart: A. Sarajedini, *IBVS* 2587 (1984).
Light curve: A. Sarajedini, *IBVS* 2587 (1984); B. Hidayat, S. P. Wiramhardya, *AsAp* **65** (1978) 143.
Spectroscopy: B. Hidayat, S. P. Wiramhardya, *AsAp* **65** (1978) 143 – phot, iden.
Identification: from Sarajedini's finding chart, the published precise positions and Sky Atlas plate ESO QB 523, taken 1977 August 17 (during decline).
Classification: nova; maximum missed; high scatter in the light curve.

V4027 Sgr **N**

(N Sgr 1968)

Discovered by D. J. MacConnell on objective prism plates taken with the Cerro Tololo Schmidt telescope, 1968 May 17 and July 3. The first plate shows a continuum with emission lines characteristic of the Orion or early nebular stage (*IBVS* 1340 (1977)).

Position: 17 59 18.87 − 28 45 23.8 (D. J. MacConnell, *IBVS* 1340 (1977))
2.026 − 3.095 (G.C.)

Range: 11.0r (8.5) – [21j LCT: ? t_3: ?

Finding chart: A. Sarajedini, *IBVS* 2587 (1984).
Light curve: A. Sarajedini, *IBVS* 2587 (1984).
Spectroscopy: D. J. MacConnell, *IBVS* 1340 (1977) – descr.
Identification: MacConnell's position corresponds to an empty field on the SRC atlas plate.
Classification: poorly known nova.

V4049 Sgr **N**

(N Sgr 1978)

Discovered by B. Stenholm and I. Lundström on an objective prism plate taken with the Uppsala Southern Schmidt telescope, 1978 March 8 (*IAU Circ* 3411 (1979)).

Position: 18 17 29.05 − 27 57 49.2 (SRC)
18 17 29.54 − 27 57 47.5 (M. F. McCarthy, B. M. Lasker, T. D. Kinman, *PASP* **93** (1981) 470))
4.652 − 6.217 (G.C.)

Range: 12p – 21j LCT: ? t_3: ?

Finding chart: E. P. Belserene, *IBVS* 1706 (1979); M. F. McCarthy, B. M. Lasker, T. D. Kinman, *PASP* **93** (1981) 470.
Light curve: E. P. Belserene, *IBVS* 1706 (1979), and private communication.
Spectroscopy: M. F. McCarthy, B. M. Lasker, T. D. Kinman, *PASP* **93** (1981) 470 – trac, ident, coronal lines.
Identification: from McCarthy's precise position and finding chart.
Classification: poorly known nova.

V4065 Sgr **N?**

(N Sgr 1980)

Discovered by M. Honda, Japan, 1980 October 28 (*IAU Circ* 3533).

Position:	18 16 30	– 24 45	(M. Honda, *IAU Circ* 3533 (1980))
	7.399	– 4.515	(G.C.)

Range: 9.0v – [18 LCT: ? t_3: ?

Identification: not made because of lack of finding chart or precise position. A GPO plate taken 1986 May does not show the object. Field map in Appendix.

Classification: amplitude suggests nova.

V4074 Sgr **NB or ZAND**

(AS 295(B), MHα 304-41)

Discovered as Hα emission-line object by P. W. Merrill and C. G. Burwell (*ApJ* **112** (1950) 72). A novalike outburst was recorded in 1965. The eruptive star is a visual companion to a K giant 3″ distant, PA = 190° (G. H. Herbig, D. Hoffleit, *ApJ* **202** (1975) L41).

Position:	18 12 51.90	– 30 52 14.85	(SRC; position of K giant)
	18 12 51.8	– 30 52 15	(G. H. Herbig, D. Hoffleit, *ApJ* **202** (1975) L41)
	1.593	– 6.689	(G.C.)

Range: **8.6**p – 12.3p (combined light) LCT: ? t_3: 120^d

Light curve: G. H. Herbig, D. Hoffleit, *ApJ* **202** (1975) L41.

Spectroscopy: G. H. Herbig, D. Hoffleit, *ApJ* **202** (1975) L41 – phot, descr, ident, line strengths; G. Wallerstein, *AsAp* **163** (1986) 337 – descr, ident, line strengths.

Identification: from Herbig and Hoffleit's paper. The system is a visual triple; the eruptive component, B, is 3″ S of A; a fainter component, C, is 4″ SE of B.

Classification: spectroscopic observations 10 years after outburst show coronal line spectrum; peculiar slow nova (G. H. Herbig, D. Hoffleit, *ApJ* **202** (1975) L41); symbiotic star? (G. Wallerstein, *AsAp* **163** (1986) 337).

V4077 Sgr **NB**

(N Sgr 1982)

Discovered by M. Honda, Japan, 1982 October 4 (*IAU Circ* 3733).

Position:	18 31 32.83	− 26 28 27.1	(SRC)
	18 31 32.75	− 26 28 28.0	(M. P. Candy, *IAU Circ* **3741** (1982))
	7.426	− 8.319	(G.C.)

Range: 8.0v − 22j LCT: Bb t_3: 100^d

Finding chart: A. Sarajedini, *IBVS* 2587 (1984).
Light curve: A. Sarajedini, *IBVS* 2587 (1984); T. Iijima, L. Rosino, *PASP* **95** (1983) 506.
Spectroscopy: J. R. Sowell, A. P. Cowley, *IBVS* 2283 (1983) – trac; T. Iijima, L. Rosino, *PASP* **95** (1983) 506 – ident, trac, rv; T. Mazeh, H. Netzer, G. Shaviv, H. Drechsel, J. Rahe, W. Wargau, J. C. Blades, C. Cacciari, W. Wamsteker, *AsAp* **149** (1985) 83 – trac, ident.
UV observations: T. Mazeh, H. Netzer, G. Shaviv, H. Drechsel, J. Rahe, W. Wargau, J. C. Blades, C. Cacciari, W. Wamsteker, *AsAp* **149** (1985) 83.
IR observations: H. Dinerstein, *AJ* **92** (9186) 1381 – IRAS observations.
Identification: comparison of Sky Atlas plates; on ESO R 522, taken 1984 August 22/23, in decline.
Classification: moderately fast nova.

N Sgr 1953 **N**

(NSV 10159)

Discovered by G. Haro, Tonantzintla Observatory, 1953 February 10 (*IAU Circ* 1391, *HAC* 1210)

Position:	18 02 00	− 29 54 56	(G. Haro, *IAU Circ* 1391 (1953))
	1.305	− 4.175	(G.C.)

Range: 10.5 − ? LCT: ? t_3: ?

Spectroscopy: G. Haro, *IAU Circ* 1391 (1953) – descr.
Identification: not possible because of lack of finding chart or precise position. Field map in Appendix.
Classification: poorly known nova.

N Sgr 1963

N

(NSV 09828)

Discovered by V. M. Blanco, Warner and Swasey Observatory, on objective prism plates taken 1963 July 17 and 18 (*IAU Circ* 1834).

Position: 17 51 33 − 28 41 11 (V. M. Blanco, *IAU Circ* 1834 (1963))
1.230 − 1.588 (G.C.)

Range: 13 − ? LCT: ? t_3: ?

Spectroscopy: V. M. Blanco, *IAU Circ* 1834 (1963).
Identification: not possible because of lack of finding chart or precise position. Field map in Appendix.
Classification: poorly known nova.

N Sgr 1983

N

Discovered by M. Wakuda, Japan, on films taken 1983 February 13. Maximum occurred probably in late 1982 (H. Kosai, *IAU Circ* 4119 (1985)).

Position: 18 04 44.33 − 28 49 54.5 (SRC)
18 04 44.41 − 28 49 55.0 (GPO plate, May 1986, in decline)
2.545 − 4.166 (G.C.)

Range: 9.5v − 21j LCT: ? t_3: ?

Finding chart: K. Ogura, H. Kosai, *Tokyo Astr Bull 2nd Ser* **273** (1985) 3155.
Light curve: K. Ogura, H. Kosai, *Tokyo Astr Bull 2nd Ser* **273** (1985) 3155.
Spectroscopy: K. Ogura, H. Kosai, *Tokyo Astr Bull 2nd Ser* **273** (1985) 3155 – phot.
Identification: from GPO plate.
Classification: nova.

N Sgr 1984

N:

Discovered by W. Liller, Viña del Mar, Chile, 1984 September 25 (*IAU Circ* 3995).

Position: 17 50 31.25 − 29 01 33.9 (E. P. Belserene, *IAU Circ* 3997 (1984))
0.823 − 1.567 (G.C.)

Range: 10.5v − ? LCT: ? t_3: ?

Finding chart: E. P. Belserene, private communication.
Light curve: W. Liller, *IAU Circ* 3998 (1984).
Identification: from Belserene's finding chart.
Classification: poorly known nova without spectroscopic verification.

NVS 12329 Sgr N??

Discovered by W. Fleming as an emission-line object; A. D. Thackeray (*MNASSA* **33** (1974) 131) suspects nova.

Position:			
	19 40 33.45	− 40 33 24.2	(Perth 70; CoD − 40° 13448)
	19 40 32	− 40 33	(A. D. Thackeray, *MNASSA* **33** (1974) 131)
	359.015	− 26.724	(G.C.)

Identification: not possible.
The object is close to CoD − 40° 13448 (9.0, spectral type K0); Fleming classified this star as B5p with Hβ emission, and Thackeray suspects that she found a temporary star near the position of CoD − 40° 13448. Since Fleming did not give the time or plate number in her investigation, a search for the discovery plate in the Harvard plate collection was impossible. Field map in Appendix.
Classification: existence and type of variability obscure.

T Sco N:

(N Sco 1860, N Sco No. 1)

Discovered by E. Luther and A. Auwers, Berlin Observatory, 1860 May 21 in the globular cluster NGC 6093 (M 80). The nova was not visible 1860 May 18 (*AN* **53** (1860) 293).

Position:			
	16 14 03.79	− 22 51 09.4	(A. Auwers, *AN* **114** (1886) 47)
	352.675	+ 19.462	(G.C.)

Range: 7v − ? LCT: ? t_3: 21^d

Light curve: A. Auwers, *AN* **114** (1886) 47; H. B. Sawyer, *JRAS Can* **32** (1938) 69.
Identification: not possible. Field map in Appendix.
Classification: fast nova in globular cluster. No spectroscopic information is available. The absolute magnitude at maximum is in accordance with cluster membership.

U Sco **NR**

(N Sco 1863, 1906, 1936, 1979; BD −17°4554)

Discovered by N. R. Pogson, Madras Observatory, 1863 May 20, as a 9^m star (Mem *RAS* **58** (1908) 90).

Position:	16 19 37.49	−17 45 42.9	(R. F. Webbink, *PASP* **90** (1978) 57)
	16 19 37.59	−17 45 43.1	(N. R. Pogson, *Mem RAS* **58** (1908) 90)
	357.669	+21.869	(G.C.)

Range: **8.8**v – 19.2v LCT: Ar t_3: 7^d

Finding chart:	H. L. Thomas, *HB* 912 (1940) 11; R. F. Webbink, *PASP* **90** (1978) 57.
Light curve:	H. L. Thomas, *HB* 912 (1940) 11; F. M. Bateson, *NZAS Publ.* **7** (1979) 44; H. W. Duerbeck, W. C. Seitter, *IBVS* 1738 (1980).
Spectroscopy:	B. S. Shylaya, T. P. Prabhu, *KOB Ser A*, *23* (1979) 213; H. W. Duerbeck, W. C. Seitter, *IBVS* 1738 (1980) – trac, ident; M. J. Barlow, J. D. Brodie, C. C. Brunt, D. A. Hanes, P. W. Hill, S. K. Mayo, J. E. Pringle, M. J. Ward, M. G. Watson, J. A. J. Whelan, A. J. Willis, *MN* **195** (1981) 61 – trac, ident; R. E. Williams, W. M. Sparks, J. S. Gallagher, E. P. Ney, S. G. Starrfield, J. W. Truran, *ApJ* **251** (1981) 221 – trac, ident; D. A. Hanes, *MN* **213** (1985) 443 – trac, post-outburst spectrum.
UV observations:	M. J. Barlow *et al.*; R. E. Williams *et al.*
Identification:	from Webbink's finding chart.
Classification:	recurrent nova; outbursts in 1863, 1906, 1936, 1979; recurrence noted by H. L. Thomas.

KP Sco **NA:**

(N Sco 1928, HV 4716)

Discovered by H. H. Swope on Harvard plates. The nova is [16.5 on 1928 June 16; maximum light occurred on 1928 June 21 (*HB* 870 (1929) 25).

Position:	17 40 54.90	−35 42 07.6	(SRC)
	17 40 55	−35 42 25	(H. H. Swope, *HB* 870 (1929) 25)
	354.058	− 3.297	(G.C.)

Range: **9.4**p – 21j? LCT: A? t_3: 38^d

Finding chart:	H. H. Swope, *HA* **90** (1939) 231; V. P. Tsesevich, M. S. Kazanasmas (1971).

Light curve: H. H. Swope, *HB* 870 (1929) 25; G. Cecchini, L. Gratton (1941) 145.

Identification: from Harvard plates MF 11883, taken 1928 June 20/21, and MF 11973, taken 1928 July 9/10; the exnova is probably a star of 21^m, NE of a brighter visual double star.

Classification: amplitude and light curve form indicate fast nova; no spectroscopic information is available.

V382 Sco N:

(N Sco 1901, HV 7006)

Discovered by H. S. Leavitt on Harvard plates. The object is first seen on a plate taken 1901 September 2 (H. H. Swope, *HB* 902 (1936) 5).

Position:			
	17 48 34.99	– 35 24 22.8	(SRC)
	17 48 35	– 35 24 33	(H. H. Swope, *HB* 902 (1936) 5)
	355.122	– 4.477	(G.C.)

Range: 9.5p – 22j LCT: Cb or B t_3: ?

Light curve: H. H. Swope, *HB* 902 (1936) 5.

Identification: from Harvard plate A 5622, taken 1901 September 3.

Classification: amplitude and light curve form suggest nova. No spectroscopic information is available. P. N. Kholopov (*PZv* **11** (1956) 325) suspects membership in the galactic cluster NGC 6475.

V384 Sco N:

(N Sco 1893, N Sgr 1893, HV 7104)

Discovered by H. H. Swope on Harvard plates. The nova is seen on only three plates between 1893 April 30 and May 18 (H. H. Swope, *HB* 902 (1936) 5).

Position:			
	17 58 21.54	– 35 39 28.0	(SRC)
	17 58 20	– 35 39 27	(H. H. Swope, *HB* 902 (1936))
	535.903	– 6.325	(G.C.)

Range: 12.3p (9.3) – 18.5j LCT: ? t_3: ?

Light curve: H. H. Swope, *HB* 902 (1936).

Identification: from Harvard plate B 9280, taken 1893 May 1, and B 9482, taken 1893 May 18.

Classification: amplitude suggests nova. No spectroscopic information is available.

V696 Sco NA

(N Sco 1944, Mayall's star No. 106)

Discovered by M. W. Mayall on Harvard objective prism plates. The nova was [$16^{m}.5$ on 1944 May 5, $7^{m}.5$ on May 15 (M. W. Mayall, *PA* **55** (1947) 109).

Position:	17 49 49.625	– 35 49 38.3	(SRC)
	17 49 46	– 35 49 46	(M. W. Mayall, *PA* **55** (1947) 109)
	354.887	– 4.909	(G.C.)

Range: 7.5p – 19.5j LCT: A? t_3: 9^d

Finding chart:	M. W. Mayall, *HB* 920 (1951) 35.
Light curve:	M. W. Mayall, *PA* **55** (1947) 109 = *Harv Repr* 300 (1947) 8; A. B. Solovyev, *ATs* 95 (1949) 6.
Identification:	from Harvard plate MF 32655, taken 1944 June 20/21.
Classification:	very fast nova.

V697 Sco NA

(N Sco 1941)

Discovered by M. W. Mayall on Harvard objective prism plates. The nova is first seen on 1941 March 9 and 10, its spectrum indicated a relatively late stage (M. W. Mayall, *PA* **55** (1947) 49).

Position:	17 47 57.11	– 37 24 10.0	(SRC)
	17 47 54.4	– 37 24 12	(M. W. Mayall, *PA* **55** (1947) 49)
	353.333	– 5.387	(G.C.)

Range: 10.2p (8) – 17j LCT:A t_3: $<15^d$

Light curve:	L. Campbell, *PA* **55** (1947) 49 = *Harv Repr* 300 (1947) 1; M. W. Mayall, *HB* 918 (1946) 1; C. Payne-Gaposchkin (1957) 11.
Spectroscopy:	M. W. Mayall, *HB* 918 (1946) 1.
Identification:	from Harvard plate MF 28949, taken 1941 April 24/25.
Classification:	very fast nova.

V707 Sco NA

(N Sco 1922, N Sco No. 3, HV 3681, HD 320693)

Discovered by A. J. Cannon on Harvard plates. The nova is first seen 1922 July 11 ($10^{m}.5$); maximum occurred 1922 July 17 (*HB* 773 (1922)).

Position:	17 45 03.12	– 36 36 57.0	(SRC)
	17 45 03.1	– 36 36 56	(S. I. Bailey, *HB* 773 (1922))
	353.713	– 4.486	(G.C.)

Range: 9.9p (9.6) – 20j LCT: Cb t_3: 49^d

Finding chart: H. H. Swope, *HA* **90**, 8 (1939); A. J. Cannon, M. W. Mayall, *HA* **112** (1949) 202.
Light curve: A. J. Cannon, *HB* 775 (1922); C. Payne-Gaposchkin (1957) 12.
Spectroscopy: A. J. Cannon, *HB* 775 (1922).
Identification: from published accurate position; only a spectral plate of large scale is available at Harvard (A 12102, taken 1922 July 30); confirmation of position is necessary.
Classification: moderately fast nova with dust formation.

V711 Sco N:

(N Sco 1906, N Sco No. 2, 19.1908 Sco, HV 3061)

Discovered by A. J. Cannon on Harvard plates. The object is seen between 1906 April 24 and 1906 October 11; maximum light occurred 1906 July 2 (A. J. Cannon, *HC* 137 (1908), *AN* **179** (1908) 7).

Position:	17 50 46.70	– 34 20 42.6	(SRC; one of a group of several stars of about 19^m)
	17 50 47	– 34 20 30	(A. J. Cannon, *HC* 137 (1908)
	356.268	– 4.325	(G.C.)

Range: **9.7**p – 19j (blend) LCT: E t_3: ?

Light curve: A. D. Walker, *HA* **84** (1923) 187; E. C. Pickering, *AN* **179** (1908) 7; G. Cecchini, L. Gratton (1941) 85; C. Payne-Gaposchkin (1957) 16.
Identification: from Harvard plates A 7779, taken 1906 June 30, and A 7784, taken 1906 July 1; crowded field, unambiguous identification is not possible. The Mira star BN Sco, 12.5p – 16p, is 4^s east, 30″ north.
Classification: very slow nova without spectroscopic information, P. N. Kholopov (*PZv* **11** (1956) 325) suspects membership in the galactic cluster NGC 6475, it is probably only an alignment.

V719 Sco NA

(N Sco 1950 No. 1)

Discovered by F. Zwicky, Mt. Palomar Observatory, 1950 July 20. Maximum light occurred on 1950 July 18; a second maximum took place on August 7 (*IAU Circ* 1280).

Position:	17 42 25.07	– 33 59 45.5	(SRC)
	17 42 24	– 33 59	(F. Zwicky, *IAU Circ* 1280 (1950))
	355.673	– 2.663	(G.C.)

Range: **9.8**p – 20.5j LCT: Cb t_3: 24^d

Light curve: E. R. Herzog, F. Zwicky, *PASP* **63** (1951) 121; K. G. Henize, D. B. McLaughlin, *AJ* **56** (1950) 74; P. F. Bok, *HB* 920 (1951) 9; C. Hoffmeister, *MVS* 124 (1950); C. Payne-Gaposchkin (1957) 11.

Spectroscopy: K. G. Henize, D. B. McLaughlin, *AJ* **56** (1950) 74 – descr; D. B. McLaughlin, *PASP* **63** (1951) 129 – descr, rv.

Identification: from Harvard plate A 27497, taken 1950 October 5/6.

Classification: fast nova with dust formation.

V720 Sco N

(N Sco 1950 No. 2)

Discovered independently by G. Haro, Tonantzintla Observatory, and E. R. Herzog, Mt. Palomar Observatory. Maximum light occurred 1950 August 7 (*IAU Circ* 1283).

Position: (1)	17 48 36.94	– 35 22 39.1	(SRC; nearest star based on position)
(2)	17 48 36.20	– 35 23 15.5	(SRC; identification on photograph)
	17 48 37.00	– 35 22 40.3	(2 outburst observations)
	355.150	– 4.469	(G.C.)

Range: 7.5p – 21 LCT: C t_3: ?

Finding chart: W. S. Krogdahl, The Astronomical Universe (MacMillan Comp., New York 1962), p. 406.

Light curve: E. R. Herzog, F. Zwicky, *PASP* **63** (1951) 121; K. G. Henize, D. B. McLaughlin, *AJ* **56** (1950) 74; P. F. Bok, *HB* 920 (1951) 9; D. J. K. O'Connell, *Pontif. Acad. Sci. Acta* **16** (1954) 49 = *Riv Repr* 8; C. Payne-Gaposchkin (1957) 11.

Spectroscopy: K. G. Henize, *AJ* **56** (1950) 74 – descr; D. B. McLaughlin, *PASP* **63** (1951) 129 – descr, rv; J. Sahade, J. Landi Dessi, *ApJ* **115** (1952) 579 – phot, ident, rv.

Identification: from Harvard plate A 27497, taken 1950 October 5/6, when the nova was already in the extremely deep minimum; identification problem.

Classification: light curve of type C with unusually rapid and deep decline. P. N. Kholopov (*PZv* **11** (1956) 325) suspects membership in the galactic cluster NGC 6475.

V721 Sco **N**

(N Sco 1950 No. 3)

Discovered by G. Haro, Tonantzintla Observatory, 1950 September 3; independent discovery by F. Zwicky, Mt. Palomar Observatory, 1950 September 7 (*IAU Circ* 1288).

Position:	17 39 09.235	– 34 39 18.15	(SRC)
	17 39 06	– 34 39	(*IAU Circ* 1288)
	354.757	– 2.438	(G.C.)

Range: 10.1 p (8) – [18.0j LCT: ? t_3: ?

Light curve: E. R. Herzog, F. Zwicky, *PASP* **63** (1951) 121; K. G. Henize, D. B. McLaughlin, *AJ* **56** (1950) 74.
Spectroscopy: K. G. Henize, D. B. McLaughlin, *AJ* **56** (1951) 74 – descr; D. B. McLaughlin, *PASP* **63** (1951) 129 – descr, rv.
Identification: from Harvard plate A 27494, taken 1950 October 5/6.
Classification: nova.

V722 Sco **NA**

(N Sco 1952 No. 1)

Discovered by G. Haro, Tonantzintla Observatory, 1952 March 10. Maximum light occurred on 1952 March 2 (*IAU Circ* 1349).

Position:	17 45 17	– 34 56 53	(*IAU Circ* 1349 (1952))
	355.168	– 3.663	(G.C.)

Range: 9.5 (9.4) – ? LCT: A t_3: 18^d

Light curve: D. Taboada, *Ton Bol* **5** (1952).
Identification: no finding chart or accurate position is available; identification is not possible. Field map in Appendix.
Classification: fast nova.

V723 Sco **NA:**

(N Sco 1952 No. 2)

Discovered by A. V. Solovyev, Stalinabad Observatory, 1952 August 11 (*IAU Circ* 1369).

Position:	(1) 17 46 40.30	– 35 23 02.9	(SRC; star 19^m, marked on chart)
	(2) 17 46 40.70	– 35 23 04.8	(SRC; star 23^m)
	17 46 44.08	– 35 23 07.2	(2 outburst observations)
	354.940	– 4.132	(G.C.)

Range: 9.8p – 19j? LCT: A t_3: 17^d

Finding chart: G. de Vaucouleurs, *ApJ* **132** (1960) 681.
Light curve: G. de Vaucouleurs, *ApJ* **132** (1960) 681; C. Payne-Gaposchkin (1957) 11.
Identification: from published finding chart.
Classification: amplitude and light curve form suggest fast nova. No spectroscopic information is available.

V728 Sco N??

(N Ara 1862, N Sco 1862)

Discovered by J. Tebbutt, New South Wales, 1862 October 4, as a 5^m star. The object was fainter than 11^m on 1862 October 13 (*MN* **38** (1878) 330).

Position:	(1) 17 35 31.22	– 45 27 06.5	(SRC; star 20^m)
	(2) 17 35 30.85	– 45 27 06.6	(SRC; star 21^m)
	17 35 31.37	– 45 27 06.6	(J. Tebbutt, *MN* **38** (1878) 330)
	345.191	– 7.566	(G.C.)

Range: 5.0 v – 20j? LCT: ? t_3: $<9^d$?

Identification: from Tebbutt's precise position; verification necessary. Two candidate stars are listed above. The object is in the vicinity of the globular cluster NGC 6388 (J. Tebbutt, *MN* **38** (1878) 330).
Classification: amplitude indicates very fast nova; the existence of the object is based on only one visual observation.

V729 Sco M: or N:

(Plaut V92)

Discovered by L. Plaut on plates taken by H. van Gent. The object is brighter than 15^m between 1936 September 12 and 21 (*Leiden Ann* **21** (1958) 3).

Position:	17 18 47.74	– 32 02 56.6	(SRC)
	17 18 48	– 32 02 39	(L. Plaut, *Leiden Ann* **21** (1958) 3)
	354.587	+ 2.544	(G.C.)

Range: 14.0p – 22j? LCT: ? t_3: ?

Finding chart: L. Plaut, *Leiden Ann* **21** (1958) 3.
Light curve: L. Plaut, *Leiden Ann* **21** (1958) 3.
Identification: from Harvard plate B 61355, taken 1936 September 17/18. The object is marked by H. H. Swope.
Classification: Mira type or nova (L. Plaut, *Leiden Ann* **21** (1958) 3). Amplitude suggests nova. No spectroscopic information is available.

V733 Sco M: or N:

(Plaut V106)

Discovered by L. Plaut on plates taken by H. van Gent. The object is brighter than 15^m between 1937 June 29 and July 20 (*Leiden Ann* **21** (1958) 3).

Position:	17 36 21.12	– 35 51 02.5	(SRC)
	17 36 21	– 35 51 18	(L. Plaut, *Leiden Ann* **21** (1958) 3)
	353.439	– 2.592	(G.C.)

Range: 13.5p – 21j? LCT: ? t_3: ?

Finding chart: L. Plaut, *Leiden Ann* **21** (1958) 3).
Light curve: L. Plaut, *Leiden Ann* **21** (1958) 3.
Identification: from Plaut's finding chart.
Classification: Mira or nova type (L. Plaut, *Leiden Ann* **21** (1958) 3). Amplitude suggests nova. No spectroscopic information is available.

V734 Sco M: or N:

(Plaut V110)

Discovered by L. Plaut on plates taken by H. van Gent. The object is brighter than 15^m between 1937 June 1 and June 21 (*Leiden Ann* **21** (1958) 3).

Position:	17 41 40.06	– 35 36 43.2	(SRC)
	17 41 41	– 35 36 52	(L. Plaut, *Leiden Ann* **21** (1958) 3)
	354.215	– 3.380	(G.C.)

Range: 14.7p – 20j LCT: ? t_3: ?

Finding chart: L. Plaut, *Leiden Ann* **21** (1958) 3.
Light curve: L. Plaut, *Leiden Ann* **21** (1958) 3.

Identification: from Plaut's finding chart.
Classification: probably long period variable, perhaps nova (L. Plaut, *Leiden Ann* **21** (1958) 3). No spectroscopic information is available.

V744 Sco N:

(Plaut V133)

Discovered by L. Plaut on plates taken by H. van Gent. The object is brighter than 15^m between 1935 February 8 and May 26; maximum light occurred on 1935 February 12 (*Leiden Ann* **21** (1958) 3).

Position:	17 50 03.93	− 31 12 58.7	(SRC)
	17 50 04	− 31 13 03	(L. Plaut, *Leiden Ann* **21** (1958) 3)
	358.889	− 2.602	(G.C.)

Range: 13.3p – 21p LCT: ? t_3: ?

Finding chart: L. Plaut, *Leiden Ann* **21** (1958) 3.
Light curve: L. Plaut, *Leiden Ann* **21** (1958) 3.
Identification: tentative identification from Plaut's finding chart.
Classification: probably nova (L. Plaut, *Leiden Ann* **21** (1958) 3). No spectroscopic information is available.

V745 Sco N:

(Plaut V139)

Discovered by L. Plaut on plates taken by H. van Gent. Maximum light occurred on 1937 May 10 (*Leiden Ann* **21** (1958) 3).

Position:	17 52 01.35	− 33 14 32.5	(SRC; tentative identification)
	17 52 04	− 33 14 02	(L. Plaut, *Leiden Ann* **21** (1958) 3)
	357.352	− 3.989	(G.C.)

Range: 11.2p – 21j LCT: A t_3: 12^d

Finding chart: L. Plaut, *Leiden Ann* **21** (1958) 3.
Light curve: L. Plaut, *Leiden Ann* **21** (1958) 3; H. W. Duerbeck, *IBVS* 2490 (1984).
Identification: from Harvard plates B 61924, B 61938 and MF 23259, taken between 1937 May 10/11 and 1937 June 3/4.
Classification: amplitude and light curve form suggest nova. No spectroscopic information is available.

V825 Sco **N**

(N Sco 1964)

Discovered by A. Przybylski, Mt. Stromlo Observatory, 1964 May 19, as a 12^{m} star several months past maximum (*IAU Circ* 1864).

Position:	17 46 35.76	− 33 31 22.3	(SRC)
	17 46 36	− 33 32 05	(A. Przybylski, *IAU Circ* 1864 (1964))
	356.531	− 3.159	(G.C.)

Range: 12p (8) – 19.0j LCT: ? t_3: ?

Finding chart: A. Przybylski, *IAU Coll. 46* (ed. F. M. Bateson *et al.*), Hamilton, NZ 1979, p. 119.
Light curve: A. Przybylski, *IAU Coll. 46* (ed. F. M. Bateson *et al.*), Hamilton, NZ 1979, p. 119.
Spectroscopy: A. Przybylski, *IAU Coll. 46* (ed. F. M. Bateson *et al.*), Hamilton, NZ 1979, p. 119 – descr, phot; K. Wilde, *PASP* **77** (1965) 208 – descr, phot.
Identification: from Przybylski's finding chart.
Classification: slow nova, maximum not observed.

V902 Sco **NB**

(N Sco 1949; KZP 7652)

Discovered independently by K. G. Henize and G. Haro. Maximum occurred near 1949 May 21 (*PASP* **73** (1951) 360).

Position:	(1) 17 22 41.20	− 39 01 31.0	(SRC; star 20.0, marked on chart)
	(2) 17 22 41.08	− 39 01 27.6	(SRC; star 21^{m}.0)
	349.294	− 2.069	(G.C.)

Range: 11p (11?) – 20.0j LCT: A? t_3: 200^{d}

Finding chart: K. G. Henize, G. Haro, *PASP* **73** (1951) 360.
Light curve: K. G. Henize, G. Haro, *PASP* **73** (1951) 360.
Spectroscopy: K. G. Henize, G. Haro, *PASP* **73** (1951) 360 – phot.
Identification: from Henize and Haro's finding chart; uncertain identification; group of stars. In obscured region of the Galaxy.
Classification: slow nova.

V916 Sco — ZAND

(N Sco 1967, SSM 1)

Discovered by N. Sanduleak, C. B. Stephenson and D. J. MacConnell on Cerro Tololo objective prism plates (*IBVS* 1376 (1978)).

Position:	17 40 32.42	− 36 02 14.9	(SRC)
	17 40 32.6	− 36 02 07	(N. Sanduleak, C. B. Stephenson, D. J. MacConnell, *IBVS* 1376 (1978))
	353.732	− 3.409	(G.C.)

Range: 14 – 17r LCT: ? t_3: ?

Finding chart: I. Lundström, B. Stenholm, *IBVS* 1393 (1978); D. A. Allen (1984).
Light curve: B. S. Carter, M. W. Feast, *IBVS* 1714 (1979).
Identification: from Allen's finding chart.
Classification: symbiotic star (D. A. Allen (1984), B. S. Carter, M. W. Feast, *IBVS* 1714 (1979)).

N Sco 1952 — N

(NSV 09663)

Discovered by G. Haro, Tonantzintla Observatory, on an objective prism plate taken 1952 April 18 (*IAU Circ* 1395, *HAC* 1174)

Position:	17 44 20.75	− 33 10 44.0	(SRC, tentative identification)
	17 43 56	− 33 10 25	(G. Haro, *IAU Circ* 1395 (1952))
	356.581	− 2.579	(G.C.)

Range: 11p – 21j? LCT: ? t_3: ?

Identification: no finding chart or precise position is available. On POSS plate in outburst? Finding chart in Appendix.
Classification: poorly known nova.

N Sco 1954 — N

(NSV 09808)

Discovered by P. Wild, Mt. Palomar Observatory, on objective prism plates taken 1954 August 30 (*IAU Circ* 1471).

Position: 17 50 27 − 30 44 57 (P. Wild, *IAU Circ* 1471 (1954))
359.333 − 2.435 (G.C.)

Range: 13.8p – ? LCT: ? t_3: ?

Identification: no finding chart or precise position is available. Identification is not possible. Field map chart in Appendix.
Classification: poorly known nova.

N Sco 1985 N

Discovered by W. Liller, Viña del Mar, Chile, 1985 September 24, when the nova was $10^m.5$. It was [12^m on 1984 September 19 (*IAU Circ* 4118).

Position: 17 53 19.01 − 31 49 14.2 (GPO plate, May 1986)
17 53 18.85 − 31 49 14.45 (SRC)
358.720 − 3.506 (G.C.)

Range: 10.5v – 20j LCT: ? t_3: ?

Finding chart: R. Lukas, *IBVS* 2852 (1986).
Spectroscopy: T. Richtler, W. Liller, *IBVS* 2871 (1986) – trac; H. W. Duerbeck, W. C. Seitter, *ApSS* **131** (1987) 467 – descr.
Identification: from GPO plate (nova in decline).
Classification: nova.

EU Sct NA

(N Sct 1949)

Discovered by C. Bertaud, Observatoire de Paris, 1949 July 31. Maximum was reached on 1949 August 5 (*IAU Circ* 1224).

Position: 18 53 34.50 − 04 16 30.4 (POSS)
18 53 34.63 − 04 16 27.7 (3 outburst observations)
29.727 − 2.980 (G.C.)

Range: **8.4**p – 18p LCT: Cb t_3: 42^d

Finding chart: Yu. N. Efremov (1961); N. E. Kurochkin, *ATs* 90-91 (1949) 2.
Light curve: L. Campbell, *Harv Repr* 327 (1949) 29; M. Beyer, *AN* **280** (1951) 273; C. Bertaud, *JO* **36** (1953) 29; M. Harwood, *Leiden Ann* **21** (1962) 404; C. Payne-Gaposchkin (1957) 12.
Spectroscopy: A. Colacevich, *ApJ* **111** (1950) 197 – ident; P. Wellmann, *ZsAp* **29** (1951) 101 – ident, rv; J. F. Heard, *JRAS Can* **47** (1953)

	109 = DDO Comm 32 – ident, rv; Ch. Fehrenbach, *CR* **229** (1949) 1059 – descr, ident, rv; Y. Andrillat, Ch. Fehrenbach, *JO* **33** (1950) 143 – phot, ident, rv.
Identification:	from published positions and finding charts.
Classification:	moderately fast nova.

FS Sct **NA:**

(N Sct 1952, N Aql 1952)

Discovered by S. Arend, Observatoire de Bruxelles, 1952 July 19. Maximum light occurred around 1952 June 23 (*IAU Circ* 1367).

Position:	18 55 36.98	– 05 28 11.4	(POSS)
	18 55 37.1	– 05 28 09	(S. Arend, *IAU Circ* 1367 (1952))
	28.895	– 3.978	(G.C.)

Range: 10.1p – 18p LCT: Bb t_3: 86^d

Finding chart:	Yu. N. Efremov (1961).
Light curve:	C. A. Whitney, *HB* 921 (1952) 27; M. Harwood, *Leiden Ann* **21** (1962) 404; F. Hunaerts, *CRO Belg No.* 54 (1953); W. H. Steavenson, *MN* **113** (1953) 258; R. Kippenhahn, *Nbl AZ* **6** (1952) 26.
Identification:	from discovery plate, communicated by H. Debehogne, Uccle.
Classification:	moderately fast nova.

FV Sct **N**

(N Sct 1960)

Discovered by M. V. Saveljeva, Sternberg Observatory Moscow, on an objective prism plate taken 1960 June 29 (*ATs* 217 (1960) 1).

Position:	18 32 02.67	– 12 57 52.5	(POSS)
	18 32 02.7	– 12 57 53.0	(S. Wyckoff, P. A. Wehinger (1978))
	19.555	– 2.252	(G.C.)

Range: 12.5p (7) – 20p LCT: ? t_3: ?

Finding chart:	N. V. Saveljeva, *ATs* 217 (1960) 1; J. J. Nassau, C. B. Stephenson, *PASP* **73** (1961) 256.
Light curve:	C. Bertaud, *IBVS* 223 (1967); P. S. Thé, *IBVS* 215 (1967); M. V. Saveljeva, *ATs* 217 (1960) 1.

Spectroscopy: J. J. Nassau, C. B. Stephenson, *PASP* **73** (1961) 256 – phot, descr.
Identification: from Nassau and Stephenson's finding chart.
Classification: nova; only the late phases of the outburst are covered by observations.

GL Sct N?

(MMO variable 1087)

Discovered by Ms. Hanner in 1954 on plates taken in 1915. The object is visible on three plates taken with the Harvard Metcalf refractor 1915 May 9 to 11 (M. Harwood, *Leiden Ann* **21** (1962) 387).

Position:	18 43 06.65	– 06 28 22.4	(POSS)
	18 43 07	– 06 28 39	(M. Harwood, *Leiden Ann* **21** (1962) 387)
	26.576	– 1.674	(G.C.)

Range: 13.6p – ? LCT: ? t_3: ?

Finding chart: M. Harwood, *Leiden Ann* **21** (1962) 387.
Light curve: M. Harwood, *Leiden Ann* **21** (1962) 387.
Identification: from Harvard plates MC 8503, taken 1915 May 9, and MC 8531, taken 1915 May 11.
Classification: small amplitude and lack of spectroscopic observations leave severe doubts concerning the nova nature.

V366 Sct N

(N Sct 1961)

Discovered by P. S. Thé on an objective prism plate taken with the Lembang Schmidt telescope, 1961 May 22 (*IBVS* 211 (1967)).

Position:	18 26 54.6	– 12 20 58	(SERC, empty field)
	18 26 56	– 12 20 45	(P. S. Thé, *IBVS* 211 (1967))
	19.516	– 0.857	(G.C.)

Range: 15.4 – [23j LCT: ? t_3: ?

Finding chart: P. S. Thé, *IBVS* 211 (1967).
Identification: sky atlas position estimated from Thé's finding chart, empty field.
Classification: poorly known nova.

V368 Sct **NA**

(N Sct 1970)

Discovered by G. E. D. Alcock, Peterborough, England, 1970 July 31 (*IAU Circ* 2269).

Position:	18 42 59.96	− 08 36 13.6	(POSS)
	18 42 59.95	− 08 36 14.5	(2 outburst observations)
	24.669	− 2.629	(G.C.)

Range: **6.9**v − 19.0p LCT: B? t_3: 30^d

Finding chart: L. J. Robinson, M. Harwood, *IBVS* 472 (1970); H. Kosai, *Tokyo Astr Bull 2nd Ser* **214** (1971) 2515; J. Cohen, A. J. Rosenthal, *ApJ* **268** (1983) 689.
Light curve: W. H. Warren, *PASP* **83** (1971) 14; F. Ciatti, L. Rosino, *AsAp* **16** (1974) 305; I. D. Howarth, *JBAA* **88** (1978) 180.
Spectroscopy: F. Ciatti, L. Rosino, *AsAp Suppl* **16** (1974) 305 – phot, ident; S. Kikuchi, J. Smolinski, *AA* **25** (1975) 305 – ident, rv.
Radio observations: V. Herrero, R. M. Hjellming, C. M. Wade, *ApJ* **166** (1971) L19.
Identification: the finding charts by Robinson and Harwood and by Kosai are incorrect. The identification was made from the published positions.
Classification: moderately fast nova.

V373 Sct **NA**

(N Sct 1975)

Discovered by P. Wild, Observatorium Zimmerwald, 1975 June 15. Maximum light occurred on 1975 May 11 (*IAU Circ* 2791).

Position:	18 52 44.18	− 07 46 59.8	(POSS)
	18 52 44.4	− 07 47 00.4	(3 outburst observations)
	26.504	− 4.397	(G.C.)

Range: **7.1**v − 18.5p LCT: Bb t_3: 85^d

Finding chart: M. I. Raff, J. Thorstensen, *PASP* **87** (1975) 593; S. Wyckoff, P. A. Wehinger (1978) 557.
Light curve: J. A. Mattei, *JRAS Can* **69** (1975) 319; L. Rosino, *ApSS* **55** (1978) 383.
Spectroscopy: L. Rosino, *ApSS* **55** (1978) 383 – trac, ident, rv; J. S. Gallagher, *ApJ* **221** (1978) 211 – trac, ident; Ch. Fehrenbach, Y. Andrillat, *CR* **281** (1975) 169.

Identification: from Harvard plates MC 39283 and MC 39284, taken 1975 June 20/21.
Classification: moderately fast nova.

V427 Sct **M?**

(N Sct 1958, SVS 7917, S-WS 12)

Discovered by S. Apriamashvili, Abastumani Observatory, as an Hα emission object on an objective prism plate taken 1958 September 17 (*ATs* 229 (1962) 1).

Position:	18 40 20.88	– 04 31 21.2	(SRC)
	18 40 21	– 04 32 30	(C. B. Stephenson, *IBVS* 966 (1975))
	27.990	– 0.165	(G.C.)

Range: 15p – 17: LCT: M? t_3: –

Finding chart: C. B. Stephenson, *IBVS* 966 (1975).
Spectroscopy: C. B. Stephenson, *IBVS* 966 (1975); S. Apriamashvili, *ATs* 229 (1962) 1; H. W. Duerbeck (unpublished).
Identification: from Stephenson's finding chart. Both the positions given by Apriamashvili and by Stephenson are incorrect.
Classification: S star, probably with Mira variability. Maxima in 1947, 1952, 1955; minima in 1951, 1952, 1962 (C. B. Stephenson, *IBVS* 966 (1975)).

NSV 11561 Sct **N??**

(MMO variable 520; variable No. 238)

Discovered by M. Harwood on Harvard plates taken 1938 August (*Leiden Ann* **21** (1962) 287).

Position:	18 53 50.635	– 08 39 29.4	(POSS)
	18 53 56	– 08 39 24	(M. Harwood, *Leiden Ann* **21** (1962) 387)
	25.847	– 5.039	1G.C.)

Range: 16.2p – 17.0p LCT: ? t_3: –

Finding chart: M. Harwood, *Leiden Ann* **21** (1962) 387.
Light curve: M. Harwood, *Leiden Ann* **21** (1962) 387.
Identification: from Harvard plates A 20355, taken 1938 August 2/3, and A 20359, taken 1938 August 3/4.
Classification: amplitude $0^m.8$; nova classification extremely unlikely. No spectroscopic evidence is available.

N Sct 1981 N??

Discovered by D. Branchett, England, 1981 January 18, as an 8^m star (*IAU Circ* 3566).

Position: 18 44 11.7 – 04 59 56 (R. Argyle, E. Clements, *IAU Circ* 3566 (1981))
28.010 – 1.235 (G.C.)

Range: 8 – [22? LCT: ? t_3: ?

Classification: many observers could not verify the object. The plate taken with the Greenwich astrograph shows a stellar object; the SERC equatorial survey yields an empty field. The existence of the object is not beyond doubt.

X Ser NB:

(N Ser 1903, 117.1908 Ser, HV 3137)

Discovered by H. Leavitt on Harvard plates. The object was near maximum light between 1903 May and September (*HC* 142 (1908), *AN* **179** (1908) 159).

Position: 16 16 41.32 – 02 22 17.8 (POSS)
16 16 41.36 – 02 22 17.6 (3 recent observations)
10.841 + 31.873 (G.C.)

Range: **8.9**p – 18.3p LCT: D t_3: ?

Finding chart: S. Wyckoff, P. A. Wehinger (1978); G. Williams (1983); A. Sh. Khatisov (1971); T. D. Kinman, C. A. Wirtanen, K. A. Janes, *ApJ Suppl* **11** (1965) 223.
Light curve: A. D. Walker, *HA* **84** (1923) 189; E. Hughes Boyce, *HA* **109** (1942) 10; G. P. Sacharov, *PZv* **10** (1954) 36; G. Cecchini, L. Gratton (1941) 72, 73; C. Payne-Gaposchkin (1957) 15.
Spectroscopy: H. W. Duerbeck, W. C. Seitter, *ApSS* **131** (1987) 467 – minimum spectrum, descr; G. Williams (1983), minimum spectrum, trac.
Identification: from published finding charts.
Classification: very slow nova without spectroscopic confirmation during outburst; after 1932 brightness fluctuations between $14^m.5$ and $16^m.2$ and a period of 275^d were observed.

RT Ser **ZAND or NC**

(N Ser 1909, 7.1917 Ser)

Discovered independently by M. Wolf (*AN* **204** (1917) 293) and E. E. Barnard (*AJ* **32** (1919) 48). The brightness increased from 1909 to 1915; maximum light occurred between 1915 and 1925.

Position: 17 37 04.10 – 11 55 03.95 (POSS)
17 37 04.10 – 11 55 04.3 (A. Sh. Khatisov (1971))
13.895 + 9.971 (G.C.)

Range: **10.6**p – 16p LCT: E t_3: ?

Finding chart: Yu. N. Efremov (1961); A. Sh. Khatisov (1971); D. A. Allen (1984).
Light curve: D. Hoffleit, *HB* 911 (1939) 41; C. Payne-Gaposchkin, S. Gaposchkin, Variable Stars, Cambridge 1938, p. 263; G. Cecchini, L. Gratton (1941) 85.
Spectroscopy: W. S. Adams, A. H. Joy, *PASP* **40** (1928) 252 – descr, rv; A. H. Joy, *PASP* **43** (1931) 353 – descr, rv; P. Swings, O. Struve, *ApJ* **92** (1940) 295 – ident, rv, line intensities; P. Swings, O. Struve, *ApJ* **96** (1942) 468 – ident, line intensities; J. Grandjean, *AAp* **15** (1952) 7 – ident, rv, line intensities; J. W. Fried, *AsAp* **81** (1980) 182 – ident, rv, line intensities.
Identification: from published finding charts.
Classification: symbiotic star; sometimes classified as extremely slow nova.

CT Ser **N**

(N Ser 1948)

Discovered by R. Bartaya, Abastumani Observatory, 1948 April 9. The nova was already 8^m on a Harvard plate taken 1948 February 6; maximum light occurred probably near the end of 1947 (*IAU Circ* 1150).

Position: 15 43 19.58 + 14 31 51.2 (POSS, near plate edge)
15 43 19.76 + 14 31 50.3 (8 outburst observations)
24.482 + 47.563 (G.C.)

Range: 7.9v (5) – 16.6p LCT: ? t_3: ?

Finding chart: A. Sh. Khatisov (1971); S. Wyckoff, P. A. Wehinger (1978).

Light curve: P. Ahnert, *Nbl AZ* **2** (1948) 18; J. Gossner, *PASP* **60** (1948) 329; A. V. Solovyev, *Tadj Tsirk* 76 (1949) 1, 69; W. Lohmann, *AN* **277** (1949) 37; M. Beyer, *AN* **280** (1951) 273; M. S. Davis, *AJ* **55** (1951) 126; R. A. Bartaya, *Abast Bull* **15** (1953) 17; J. Ashbrook *AJ* **58** (1953) 176.

Spectroscopy: O. J. Wilson, *PASP* **60** (1948) 327 – ident; M. Bloch, *AAp* **13** (1950) 390 – trac, phot, ident, rv; R. A. Bartaya, *Abast Bull* **15** (1953) 17 – phot, ident; M. Bloch, *CR* **227** (1948) 333 – phot, ident.

Identification: from Wyckoff and Wehinger's finding chart; on POSS in decline.

Classification: nova whose late phases are well studied.

DZ Ser **N**

(N Ser 1960)

Discovered by J. J. Nassau and C. B. Stephenson, Warner and Swasey Observatory, 1960 July 29 (*HAC* 1526).

Position:	17 58 12.72	– 10 33 50.35	(POSS)
	17 58 12.8	– 10 34 00.5	(S. Wyckoff, P. A. Wehinger (1978))
	17.708	+ 6.169	(G.C.)

Range: 14.0p (7?) – 21p LCT: ? t_3: ?

Finding chart: J. J. Nassau, C. B. Stephenson, *PASP* **93** (1961) 256.

Spectroscopy: J. J. Nassau, C. B. Stephenson, *PASP* **93** (1961) 256.

Identification: from Nassau and Stephenson's finding chart.

Classification: poorly known nova.

FH Ser **NA**

(N Ser 1970)

Discovered by M. Honda, Japan, 1970 February 13 (*IAU Circ* 2212).

Position:	18 28 16.24	+ 02 34 42.8	(SRC)
	18 28 16.30	+ 02 34 42.2	(6 outburst observations)
	32.909	+ 5.786	(G.C.)

Range: **4.5**v – 16.2p LCT: Cb t_3: 62^d

Finding chart: A. Sh. Khatisov (1971); G. Williams (1983); M. Burkhead, M. A. Seeds, *ApJ* **160** (1970) L51; H. Kosai, *Tokyo Astr Bull 2nd Ser* **214**

(1971) 2515; L. Rosino, F. Ciatti, M. della Valle, *AsAp* **158** (1986) 34.

Light curve: E. F. Borra, P. H. Andersen, *PASP* **82** (1970) 1070; E. F. Borra, *PASP* **83** (1971) 447; M. S. Burkhead, W. S. Penhallow, R. K. Honeycutt, *PASP* **83** (1971) 338; L. Rosino, F. Ciatti, M. della Valle, *AsAp* **158** (1986) 34.

Spectroscopy: P. H. Andersen, E. F. Borra, O. V. Dubas, *PASP* **83** (1971) 5 – trac, phot, ident, rv; J. Grygar, J. Smolinski, J. B. Hutchings, *PASP* **83** (1971) 15 – trac, ident, rv; N. R. Walborn, *PASP* **83** (1971) 813 – phot, descr; J. B. Hutchings, J. Smolinski, J. Grygar, *Victoria Publ* **14** (1972) 17 – ident, rv, trac; L. Rosino, F. Ciatti, M. della Valle, *AsAp* **158** (1986) 34 – ident, trac.
G. Williams (1983) – minimum spectrum, trac; H. W. Duerbeck, W. C. Seitter, *ApSS* **131** (1987) 467 – minimum spectrum, descr.

UV observations: J. S. Gallagher, A. D. Code, *ApJ* **189** (1974) 303.

IR observations: A. R. Hyland, G. Neugebauer, *ApJ* **160** (1970) L177; R. M. Mitchell, G. Robinson, A. R. Hyland, G. Neugebauer, *MN* **216** (1985) 1057; H. Dinerstein, *AJ* **92** (1986) 1381 – IRAS observations.

Radio observations: C. M. Wade, R. M. Hjellming, *ApJ* **163** (1971) L65; E. R. Seaquist, J. Palimaka, *ApJ* **217** (1977) 781; R. M. Hjellming, C. M. Wade, N. R. Vandenberg, R. T. Newell, *AJ* **84** (1979) 1619.

Identification: from Burkhead and Seeds' finding chart.

Classification: moderately fast nova with dust formation and good coverage in all wavelengths regions; it constitutes a verification of the constant luminosity model (J. S. Gallagher, *AJ* **82** (1977) 209).

LW Ser **NA**

(N Ser 1978)

Discovered by M. Honda, Japan, 1978 March 5 (*IAU Circ* 3186)

Position:	17 48 59.66	– 14 43 08.7	(SRC – in decline)
	17 48 59.74	– 14 43 08.2	(Y. Kosai, H. Kosai, K. Hamajima, *IAU Circ* 3188 (1978))
	12.959	+ 6.047	(G.C.)

Range: **8.3**v – 21p LCT: Cb t_3: 50^d

Light curve: J. A. Mattei, *JRAS Can* **73** (1979) 50.

Spectroscopy: T. P. Prabhu, G. C. Anupama, *ApSS* (in press) – trac, spectrophotometry.

IR observations: P. Szkody, H. M. Dyck, R. W. Capps, E. E. Becklin, D. P.

	Cruikshank, *AJ* **84** (1979) 1359; R. D. Gehrz, G. L. Grasdalen, J. A. Hackwell, *ApJ* **237** (1980) 855.
Identification:	comparison of POSS and SRC equatorial survey; on POSS at limit.
Classification:	nova with dust formation.

MU Ser **NA**

(N Ser 1983)

Discovered by M. Wakuda, Japan, 1983 February 22 (*IAU Circ* 3777).

Position:	17 53 02.42	– 14 00 52.9	(SRC – in decline)
	17 53 02.51	– 14 00 52.0	(H. Kosai, *IAU Circ* 3788 (1983))
	14.092	+ 5.569	(G.C.)

Range: **7.7**v – [21p ([19r) LCT: A? t_3: 5^d

Light curve:	E. M. Schlegel, R. K. Honeycutt, R. H. Khaitchuk, *PASP* **97** (1985) 1075.
Spectroscopy:	T. Iijima, S. Ortolani, L. Rosino, *IAU Circ* 3790 (1983); E. M. Schlegel, R. K. Honeycutt, R. H. Khaitchuk, *PASP* **97** (1985) 1075.
Identification:	comparison of POSS and SRC equatorial survey; empty field on POSS.
Classification:	very fast nova.

XX Tau **NA**

(N Tau 1927, 100.1927 Tau)

Discovered by A. Schwassmann and A. A. Wachmann, Hamburger Sternwarte, on an objective prism plate taken 1927 November 18. Harvard plates show that the maximum occurred 1927 October 1 (*BZ* **9** (1927) 82).

Position:	05 16 31.05	+ 16 39 57.9	(POSS)
	05 16 31.19	+ 16 39 58.3	(4 outburst observations)
	187.104	– 11.653	(G.C.)

Range: **5.9**p – 18.5p LCT: Cb t_3: 42^d

Finding chart:	A. Schwassmann, A. A. Wachmann, *AN* **232** (1928) 272.
Light curve:	A. J. Cannon, *BZ* **9** (1927) 85; A. A. Wachmann, *AN* **232** (1929) 409; M. Beyer, *AN* **235** (1929) 427; G. Cecchini, L. Gratton (1941) 140–142; C. Payne-Gaposchkin (1957) 12.

Spectroscopy: A. Schwassmann, A. A. Wachmann, *AN* **232** (1928) 272 – phot, descr.
Identification: from Harvard plates I 46454, taken 1927 November 29/30, and I 46527, taken 1928 January 17/18.
Classification: moderately fast nova with dust formation.

RR Tel — NC or ZAND

(166.1908 Tel, HV 3181)

The variability was discovered by W. Fleming on Harvard plates (*HC* 143 (1908), *AN* **179** (1908) 191). The nova outburst was noted by P. Kirchoff and R. P. de Kock (*MNASSA* **7** (1948) 74).

Position:			
	20 00 20.13	– 55 52 03.2	(SRC)
	20 00 18	– 55 51 45	(W. Fleming, *AN* **179** (1908) 191)
	342.163	– 32.242	(G.C.)

Range: **6.8**p – 12.5 ... 16p LCT: E t_3: $>2000^d$

Finding chart: M. W. Mayall, *HB* 919 (1949) 15.
Light curve: M. W. Mayall, *HB* 919 (1949) 15; A. Heck, R. Viotti, *AsAp* **142** (1985) 515; S. J. Kenyon (1986) 242; pre-outburst: E. L. Robinson, *AJ* **80** (1975) 341.
Spectroscopy: A. D. Thackeray, *MN* **110** (1950) 46 – phot, descr; A. D. Thackeray, *MN* **113** (1953) 211 – phot, ident, rv; A. D. Thackeray, *MN* **115** (1956) 236 – phot, ident; A. D. Thackeray, B. L. Webster, *MN* **168** (1974) 101 – trac, ident; A. D. Thackeray, *Mem RAS* **83** (1977) 1 – trac, ident; S. R. Pottasch, C. M. Varsavsky, *AsAp* **23** (1960) 516 – ident, rv, line intensities; M. Friedjung, *MN* **133** (1966) 401 – rv, line intensities; L. H. Aller, R. S. Polidan, E. J. Rhodes, jr., G. W. Wares, *ApSS* **20** (1973) 93 – ident, line intensities; L. H. Aller, C. D. Keyes, *ApSS* **30** (1974) 287 – ident, line intensities.
UV observations: M. V. Penston, P. Benvenuti, A. Casstella, A. Heck, P. L. Selvelli, F. Macchetto, D. Ponz, C. Jordan, N. Cramer, F. Rufener, J. Manfroid, *MN* **202** (1983) 833; M. A. Hayes, H. Nussbaumer, *AsAp* **161** (1986) 287.
IR observations: M. W. Feast, P. A. Whitelock, R. M. Catchpole, G. Roberts, B. S. Carter, *MN* **202** (1983) 951.
Identification: from M. W. Mayall's finding chart.
Classification: symbiotic star. Before outburst, it was known as a long period variable with $t_{max} = 2\,430\,900 + 386.73 \cdot E$.

UW Tri **N:**

(N Tri 1983)

Discovered by N. E. Kurochkin, Moscow, 1983 September 11 (*IAU Circ* 3869).

Position:	02 42 14.60	+ 33 18 48.6	(R. W. Argyle, *IAU Circ* 3878 (1983))
	148.627	− 23.605	(G.C.)

Range: 15p − [21p LCT: ? t_3: ?

Light curve:	N. E. Kurochkin, *IAU Circ* 3869 (1983).
Identification:	from Argyle's position; empty field on POSS.
Classification:	poorly known nova; no spectroscopic observations are available.

N Tri 1853 **N??**

(BD + 34°620, NSV 00856)

BD observations by E. Schönfeld showed a star of $9^{m}.5$ on 1853 September 30 and 1856 October 30. The object was invisible in 1904 and later (E. C. Pickering, *HA* **70** (1909) 93, 211).

Position:	03 15 57.03	+ 35 18 43.6	(POSS, see remark)
	03 15 56.9	+ 35 17 38	(BD)
	153.944	− 18.325	(G.C.)

Range: 9.5v − ? LCT: ? t_3: ?

Identification:	the POSS chart shows a nonstellar object, probably a galaxy with bright nucleus, near the BD position. The long visibility (1853–1856) is unusual for a nova, but not for an active galaxy. Additional observations are needed.
Classification:	an object whose existence is not established beyond doubt.

KY TrA **XND**

(N TrA 1974, TrA X-1, A1224-61)

Discovered as an X-ray source by the Ariel-5 satellite. The optical identification was made by P. Murdin *et al.* (*MN* **178** (1977) 27p).

Position:	15 24 05.52	− 61 42 34.5	(SRC)
	15 24 05.3	− 61 42 35	(P. Murdin *et al.*, *MN* **178** (1977) 27p)
	320.319	− 4.427	(G.C.)

Range: 17.5B – 22j LCT: ? t_3: 450^d

Finding chart: P. Murdin, R. E. Griffiths, K. A. Pounds, M. G. Watson, A. J. Longmore, *MN* **178** (1977) 27p.
Light curve: P. Murdin *et al.*, *MN* **178** (1977) 27p.
Identification: from finding chart by Murdin *et al.*
Classification: X-ray nova.

Nova TrA (?) ZAND

(Hen 1242, He 2-177, PN 326-10.1, Cn 1-2)

First noted by A. J. Cannon as a peculiar emission-line star (*HC* 224 (1921)).

Position: 16 40 00.03 – 62 31 40.6 (SRC)
16 40 05 – 62 33 12 (K. G. Henize, *ApJ Suppl* **14** (1967) 125)
326.414 – 10.939 (G.C.)

Range: ? LCT: ? t_3: ?

Finding chart: L. Perek, L. Kohoutek (1967); D. A. Allen (1984).
Spectroscopy: A. J. Cannon, *HC* 224 (1921); K. G. Henize, *ApJ Suppl* **14** (1967) 125 – descr; L. Webster, *PASP* **78** (1966) 136 – descr; E. D. Carlsson, K. G. Henize, *ApJ* **188** (1974) L47 – ident.
Identification: from published finding charts.
Classification: slow nova, possibly X-ray source (2U1639-62, 3U1632-64) (E. D. Carlson, K. G. Henize, *ApJ* **188** (1974) L47); symbiotic star (D. A. Allen (1984)).

RW UMi NB

(N UMi 1956, SVS 1359)

Discovered by V. Satyvaldiev, Dushanhe, in 1962 (*ATs* 232 (1962); *IBVS* 18 (1962)).

Position: 16 49 55.77 + 77 07 16.2 (POSS; at limit)
16 49 48 + 77 07 20 (B. V. Kukarkin, *IBVS* 18 (1962))
109.638 + 33.152 (G.C.)

Range: **6**p – 21p (19p after outburst) LCT: Ba t_3: 140^d

Finding chart: F. Zwicky, *KVB* **4** (1965) 169; P. Ahnert, *MVS* 732 (1963); B. V. Kukarkin, *IBVS* 18 (1962); V. Satyvaldiev, *Tadj Bull* **36** (1963) 37.

Light curve: V. Satyvaldiev, *Tadj Bull* **36** (1963) 37; P. Ahnert, *MVS* 731–733 (1963).
Spectroscopy: F. Zwicky, *KVB* **4** (1965) 169 – trac, ident.
Identification: from Zwicky's finding chart and a recent CCD frame.
Classification: slow nova at high galactic latitude.

CN Vel NB

(N Vel 1905, N Vel No. 1, 154.1906 Vel, HV 1268, HD 95821)

Discovered by H. Leavitt on Harvard plates. The nova is visible from 1905 December 5 to 1906 June (*HC* 121 (1906), *AN* **173** (1907) 295).

Position:			
(1)	11 00 28.30	– 54 06 59.6	(SRC; brightest star, 17^m)
(2)	11 00 27.965	– 54 06 59.6	(SRC, faint W component)
(3)	11 00 28.50	– 54 06 55.9	(SRC, faint NE component)
	11 00 30	– 54 07 01	(H. Leavitt, *HC* 121 (1906))
	287.431	+ 5.179	(G.C.)

Range: 10.2p – 17p? LCT: D t_3: $>800^d$

Light curve: H. Leavitt, *HC* 121 (1906); H. E. Wood, *UOC* 48 (1920) 52; A. D. Walker, *HA* **84** (1923) 189; G. Cecchini, L. Gratton (1941) 81, 82.
Spectroscopy: A. J. Cannon, *HA* **76** (1916) 19 – descr.
Identification: from Harvard plates A 8224, taken 1907 March 29, and A 8317, taken 1907 May 29.
Classification: very slow nova.

CQ Vel NA:

(N Vel 1940)

Discovered by C. J. van Houten, Leiden, on Franklin-Adams plates. Maximum light occurred around 1940 April 19 (*Leiden Ann* **20** (1950) 7).

Position:			
(1)	08 57 21.06	– 53 08 35.2	(SRC)
(2)	08 57 21.26	– 53 08 34.3	(SRC; faint star in vicinity)
	08 57 25	– 53 08 56	(C. J. van Houten, *Leiden Ann* **20** (1950) 7)
	272.333	– 4.895	(G.C.)

Range: 9.0p (8.9) – 21j LCT: Cb t_3: 50^d

Finding chart: C. J. van Houten, *Leiden Ann* **20** (1950) 7.
Light curve: D. Hoffleit, *AJ* **55** (1950) 149; C. Payne-Gaposchkin (1957) 12.

Identification:	from Harvard plates B 65262, taken 1940 April 29/30, and B 66110, taken 1940 December 24/25.
Classification:	moderately fast nova with possible dust formation.

N Vir 1929 N:

(NSV 06201, 378.1931 Vir)

Discovered by H. Schneller, Babelsberg Observatory, on two patrol plates taken 1929 February 1 and 3; not seen on 1929 February 9 and later (*AN* **243** (1931) 335).

Position:			
	13 18 30.17	+02 09 11.5	(POSS)
	13 18 28	+02 09 23	(H. Schneller, *AN* **243** (1931) 335)
	319.882	+63.784	(G.C.)

Range: 11p – 19p LCT: ? t_3: ?

Finding chart:	H. Schneller, *AN* **243** (1931) 335.
Light curve:	H. Duerbeck, *IBVS* 2502 (1984).
Identification:	from Harvard plates AC 27289, taken 1929 February 4/5, and RH 982, taken 1929 February 7/8.
Classification:	poorly known object without spectroscopic confirmation. The large amplitude makes nova not unlikely.

CK Vul N:

(N Vul 1670, N Vul No. 1, 11 Vul)

Discovered by D. Anthelme, Dijon, 1670 June 20.

Position:			
	19 45 34.97	+27 11 10.6	(M. Shara, A. F. J. Moffat, R. F. Webbink, *ApJ* **294** (1985) 271))
	63.381	+ 0.989	(G.C.)

Range: 2.6v – 20.7(?) LCT: pec (D?) t_3: ?

Finding chart:	M. Shara, A. F. J. Moffat, R. F. Webbink, *ApJ* **294** (1985) 271.
Light curve:	M. Shara, A. F. J. Moffat, R. F. Webbink, *ApJ* **294** (1985) 271.
Nebular shell:	M. Shara, A. F. J. Moffat, R. F. Webbink, *ApJ* **294** (1985) 271 – photography and spectroscopy.
Identification:	from Shara *et al.*'s finding chart. First recovered by M. Shara and A. F. J. Moffat, *ApJ* **258** (1982) L41. A. A. Wachmann's identification (*KVB* **34** (1962) 119) is incorrect (M. F. Walker, *PASP* **75** (1963) 458).
Classification:	slow nova with peculiar light curve and peculiar, planetary nebula-like remnant.

LU Vul NA

(N Vul 1968 No. 2)

Discovered by L. Kohoutek, Hamburger Sternwarte, 1968 October 14. The rise to maximum took place on 1968 July 16/17; maximum light occurred July 20/21 (*IAU Circ* 2106, 2109).

Position:	19 43 34.18	+ 28 28 07.6	(L. Kohoutek, W. Dieckvoss, *IAU Circ* 2108 (1968))
	64.261	+ 2.022	(G.C.)

Range: **9.5**p – [21p LCT: Ao t_3: 21^d

Finding chart: L. Rosino, G. Chincarini, A. Mammano, *ApSS* **4** (1969) 392; A. Terzan, *AsAp* **2** (1969) 100.

Light curve: L. Rosino, G. Chincarini, A. Mammano, *ApSS* **4** (1969) 392; A. Terzan, *AsAp* **2** (1969) 100.

Spectroscopy: L. Rosino, G. Chincarini, A. Mammano, *ApSS* **4** (1969) 392 – phot, trac, ident; F. M. Stienon, *PASP* **81** (1969) 613 – phot, descr.

Identification: a star of 19.5 is close to the position of the nova, but does not coincide exactly with it; the prenova is probably fainter than 21p (*IAU Circ* 2111 (1968)).

Classification: fast nova.

LV Vul NA

(N Vul 1968 No. 1)

Discovered by G. E. D. Alcock, Peterborough, England, 1968 April 15 (*IAU Circ* 2066).

Position:	19 45 57.37	+ 27 02 48.4	(POSS)
	19 45 57.48	+ 27 02 48.7	(5 outburst observations)
	63.303	+ 0.846	(G.C.)

Range: **5.17**v – 16.9p LCT: Ba t_3: 37^d

Finding chart: A. Terzan, *AsAp* **2** (1969) 100; I. Meinunger, *IBVS* 272 (1968); T. Z. Dvorak, M. Winiarski, *AA* **22** (1972) 33; A. Sh. Khatisov (1971).

Light curve: J. D. Fernie, *PASP* **81** (1969) 374; J. Dorschner, C. Friedemann, W. Pfau, *AN* **291** (1969) 217; J. Grygar, L. Kohoutek, *BAC* **20** (1969) 226; J. Isles, *JBAA* **83** (1972) 44; A. Terzan, M. Bally, A. Durand, *AsAp* **18** (1972) 471; P. Tempesti, *AsAp* **20** (1972) 63; C. Bartolini,

	P. Battistini, C. delli Ponti, A. Guarneri, *Mem SA It* **40** (1969) 529; pre-outburst: E. L. Robinson, *AJ* **80** (1975) 515.
Spectroscopy:	J. Dorschner, C. Friedemann, W. Pfau, *AN* **291** (1969) 217 – ident, rv; M. Bloch, *CR Sér B* **268** (1969) 106 – phot, ident, rv; J. B. Hutchings, *PASP* **82** (1970) 603 – trac, rv; J. Grygar, M. Sobotka, S. Stefl, *BAC* **32** (1981) 88 – ident, rv; Y. Andrillat, L. I. Antipova, M. B. Babaev, *AZh* **63** (1986) 128 = *Sov Astr* **30** (1986) 79 – trac, ident, rv.
Identification:	from published finding charts.
Classification:	well observed fast nova.

NQ Vul — NA

(N Vul 1976)

Discovered by G. E. D. Alcock, Peterborough, England, 1976 October 21 (*IAU Circ* 2997).

Position:	19 27 03.95	+20 21 43.6	(POSS)
	19 27 04.02	+20 21 43.45	(2 outburst observations)
	55.355	+ 1.290	(G.C.)

Range: **6.0**v – 18.5p LCT: Bb t_3: 65^d

Finding chart:	J. Cohen, A. Rosenthal, *ApJ* **268** (1983) 689.
Light curve:	Y. Yamashita, K. Ishimura, M. Kakagiri, Y. Norimoto, H. Maehara, K. Miyajima, *PASJ* **29** (1977) 527; H. W. Duerbeck, W. C. Seitter, *AsAp* **75** (1979) 297; J. A. Mattei, *JRAS Can* **72** (1978) 178.
Spectroscopy:	Ch. Fehrenbach, Y. Andrillat, *CR Sér* **B 284** (1977) 149 – phot, ident; G. Klare, B. Wolf, *AsAp Suppl* **33** (1978) 327 – phot, ident, trac, rv; Y. Yamashita, K. Ishimura, M. Hakagiri, Y. Norimoto, H. Maehara, K. Miyajima, *PASJ* **29** (1977) 527 – phot; M. J. Cottrell, S. E. Smith, *PASP* **90** (1978) 441 – phot, rv; J. W. Younger, *AJ* **85** (1980) 1232 – trac, rv; L. V. Gudim, V. G. Karetnikov, Yu. A. Medvedyev, *AZh* **59** (1982) 711 = *Sov Astr* **26** (1982) 434 – trac, descr, rv.
IR observations:	S. Sato, K. Kawara, Y. Kobayashi, T. Maihara, N. Oda, H. Okuda, T. Ijima, K. Noguchi, *PASJ* **30** (1978) 419; E. P. Ney, B. F. Hatfield, *ApJ* **219** (1978) L111; R. M. Mitchell, A. Evans, M. F. Bode, *MN* **205** (1983) 1141.
Identification:	from Harvard plate SH 5651, taken 1976 October 22/23.
Classification:	moderately fast nova.

PU Vul **ZAND or NC**

(N Vul 1979, novalike object in Vul, object Kuwano-Honda)

Discovered by M. Honda, Japan, 1978 August 21, and by Y. Kuwano, Japan, 1979 April 5 (*IAU Circ* 3344, 3348 (1979)).

Position:	20 19 01.08	+ 21 24 44.4	(POSS)
	20 19 01.09	+ 21 24 43.2	(R. W. Argyle, *IAU Circ* 3348 (1979))
	62.575	− 8.532	(G.C.)

Range: **9.0**p – 16p (var) LCT: pec (E?) t_3: ?

Finding chart: M. Honda, K. Ishida, T. Noguchi, Y. Norimoto, M. Nakagiri, T. Soyano, Y. Yamashita, *Tokyo Bull 2nd Ser* **262** (1979) 2983.

Light curve: M. H. Liller, W. Liller, *AJ* **84** (1979) 1357; W. Wenzel, *IBVS* 1608 (1979); M. Honda, K. Ishida, T. Noguchi, Y. Norimoto, M. Nakagiri, T. Soyano, Y. Yamashita, *Tokyo Bull 2nd Ser* **262** (1979) 2983; H. S. Mahra, S. C. Joshi, J. B. Srivastava, S. L. Dhir, *IBVS* 1683 (1979); M. Nakagiri, Y. Yamashita, *Tokyo Bull 2nd Ser* **263** (1980) 2993; A. Purgathofer, A. Schnell, *IBVS* 2264 (1983); J. Hron, H. M. Maitzen, *IBVS* 2273 (1983); Z. Liu, X. Hao, B. Mei, *IBVS* 2291 (1983); T. S. Belyakina, N. I. Bondar, R. E. Gershberg, Yu. S. Efimov, V. I. Krasnobabtsev, P. P. Petrov, V. Piirola, I. S. Savanov, K. K. Chuvaev, N. I. Shakovskaya, N. M. Shakovskoj, V. I. Shenavrin, *Krim Izv* **72** (1985) 3; S. J. Kenyon (1986) 246.

Spectroscopy: Y. Yamashita, H. Maehara, Y. Norimoto, *PASJ* **34** (1982) 269 – trac, descr, rv; R. E. Gershberg, V. I. Krasnobabtsev, P. P. Petrov, K. K. Chuvaev, *AZh* **59** (1982) 6 = *Sov. Astr* **26** (1982) 3 – trac, ident, rv, line strengths; T. S. Belyakina, N. I. Bondar, R. E. Gershberg, Yu. S. Efimov, V. I. Krasnobabtsev, P. P. Petrov, V. Piirola, I. S. Savanov, K. K. Chuvaev, N. I. Shakovskaya, N. M. Shakhovskoj, V. I. Shenavrin, *Krim Izv* **72** (1985) 3 – spectrophotometry, trac, ident, line strengths; S. J. Kenyon, *AJ* **91** (1976) 563 – trac, ident, rv.

Polarimetry: T. S. Belyakina, Yu. S. Efimov, E. P. Pavlenko, V. I. Shenavrin, *AZh* **59** (1982) 1 = *Sov Astr* **26** (1982) 1.

Identification: from published positions and finding chart.

Classification: extremely slow nova of RT Ser type? (T. S. Belyakina, N. I. Bondar, D. Chochol, K. K. Chuvaev, Yu. S. Efimov, R. E. Gershberg, J. Grygar, L. Ilric, V. I. Krasnobabtsev, P. P. Petrov, V. Piirola, I. S. Savanov, N. I. Shakhovskaya, N. M. Shakovskoj, V. I. Shenavrin, *AsAp* **132** (1984) L12); symbiotic star (S. J. Kenyon 1986).

PW Vul **NA**

(N Vul 1984 No. 1)

Discovered by M. Wakuda, Japan, 1984 July 27 (*IAU Circ* 3963).

Position:	19 24 03.50	+ 27 15 54.8	(GPO, May 1986)
	19 24 03.45	+ 27 15 54.4	(POSS)
	19 24 03.45	+ 27 15 54.4	(2 outburst positions)
	61.098	+ 5.197	(G.C.)

Range: **6.4**v – 17: LCT: Bb t_3: 97^d

Finding chart: H. Ridley, *JBAA* **95** (1984) 21.
Light curve: E. Schweitzer, *BAFOEV* **13** (1985) 8: R. I. Noskova, G. V. Zaitseva, E. A. Kolotikov, *Pisma AZh* **11** (1985) 613 = *Sov. Astr Lett* **11** (1985) 257; *MVS* **10** (1985) 156.
Spectroscopy: E. Schweitzer, *Astronomie* **100** (1986) 247 – phot; B. S. Shylaya, ESA Workshop 'Recent Results on Cataclysmic Variables', *ESA SP-236* (1985) 187 – trac, spectrophotometry; S. J. Kenyon, R. A. Wade, *PASP* **98** (1986) 935 – iden, trac, spectrophotometry.
X-ray observations: H. Ögelmann, K. Beuermann, J. Krautter, ESA Workshop 'Recent Results on Cataclysmic Variables', *ESA SP-236* (1985) 177.
Identification: from published precise positions.
Classification: moderately fast nova with structured light curve.

N Vul 1984 No. 2 **NA**

Discovered by P. Collins, Cardiff, California, 1984 December 27. The object was invisible on 1984 December 18 (*IAU Circ* 4023).

Position:	20 24 40.55	+ 27 40 47.2	(GPO, May 1986)
	20 24 40.565	+ 27 40 48.8	(POSS)
	20 24 40.53	+ 27 40 48.2	(A. R. Klemola, *IAU Circ* 4024 (1984))
	68.511	– 6.026	(G.C.)

Range: **5.6**v – 19p LCT: ? t_3: 40^d

Finding chart: H. Ridley, *JBAA* **95** (1985) 109; E. Schweitzer, *BAFOEV* **31** (1985) 9.
Light curve: Yu. K. Bergner, S. L. Bondarenko, A. S. Miroshnichenko, R. V. Yudin, N. Yu. Yutanov, K. S. Kuratov, D. B. Mukanov, *Pisma AZh* **11** (1985) 832 = *Sov Astr Lett* **11** (1985) 353; E. Schweitzer, *BAFOEV* **31** (1985) 9; E. Schweitzer, *Astronomie* **100** (1986) 247.

Spectroscopy: Y. Andrillat, L. Houziaux, ESA Workshop 'Recent Results on Cataclysmic Variables', *ESA SP-236* (1985) 187 – ident, rv, trac, phot.

IR observations: R. D. Gehrz, G. L. Grasdalen, J. A. Hackwell, *ApJ* **298** (1985) L47.

Identification: from published precise positions.

Classification: moderately fast nova with strong [Ne II] lines in the infrared.

THE ATLAS

Each finding chart covers a region of 4.′3 × 4.′3, unless noted otherwise. North is on the top, west to the right. In most cases, the object is centered, it is marked with two lines. The same refers to 'empty field objects' whose positions are precisely known. In the rare cases, where an approximate position yielded an empty field, it is marked with a small circle.

The finding charts labelled O or E are enlargements from blue and red POSS charts, © 1960 National Geographic Society – Palomar Sky Survey. Reproduced by permission of the California Institute of Technology (86-8).

The finding charts labelled J or F are enlargements from blue-green or red charts of the ESO/SERC Southern Sky Survey or the SERC Equatorial Survey, © Royal Observatory Edinburgh.

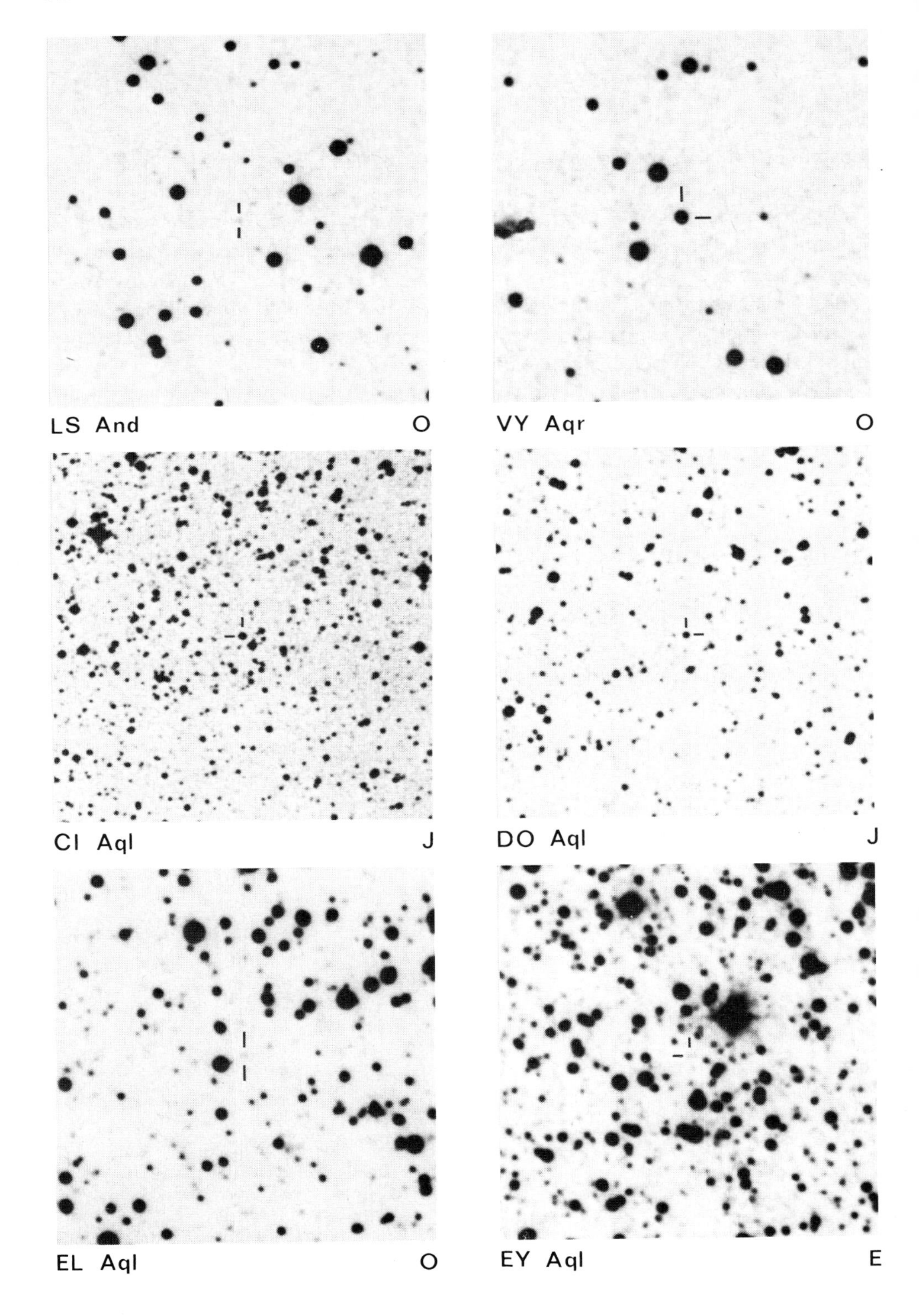
LS And
O
VY Aqr
O
CI Aql
J
DO Aql
J
EL Aql
O
EY Aql
E

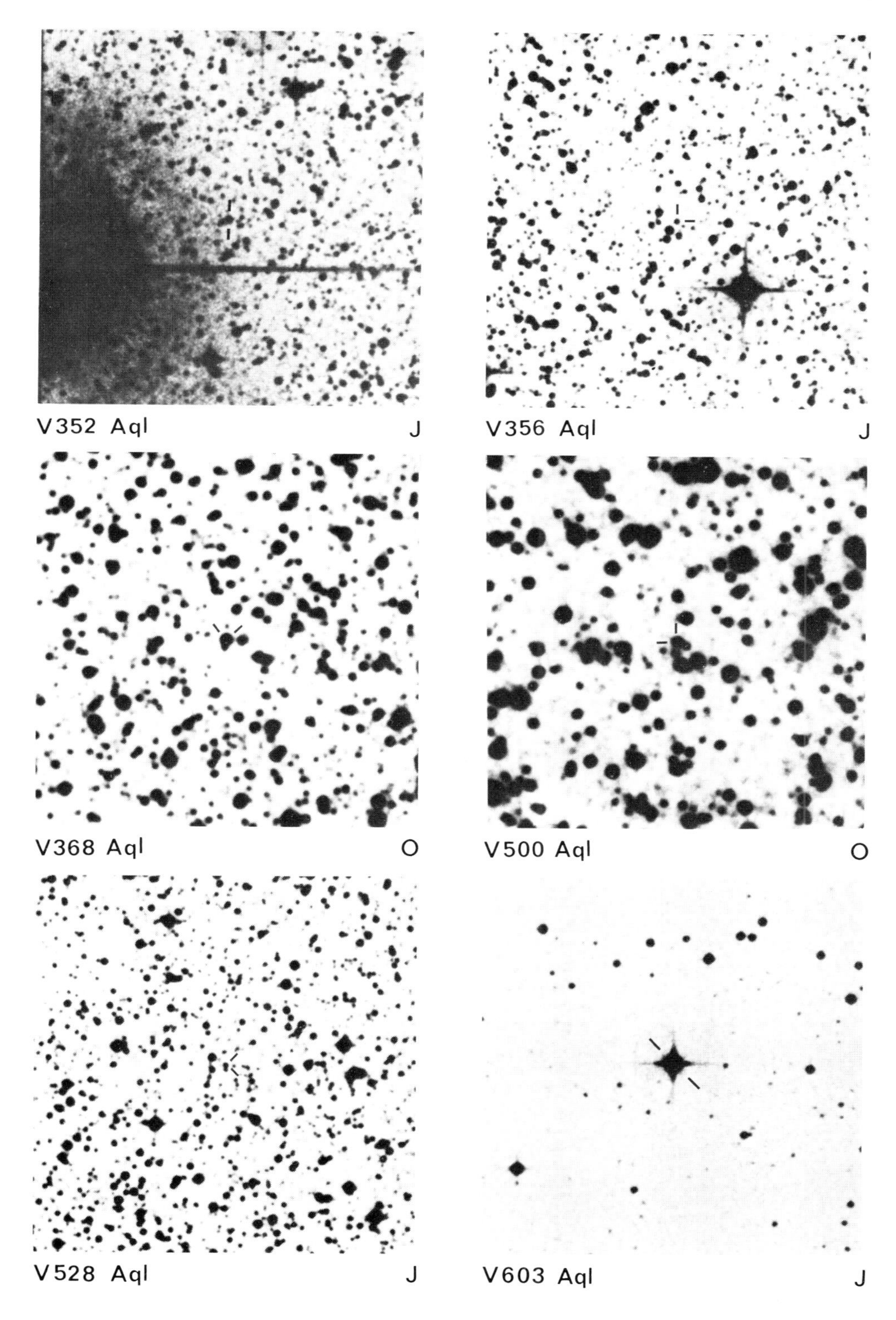

V352 Aql J

V356 Aql J

V368 Aql O

V500 Aql O

V528 Aql J

V603 Aql J

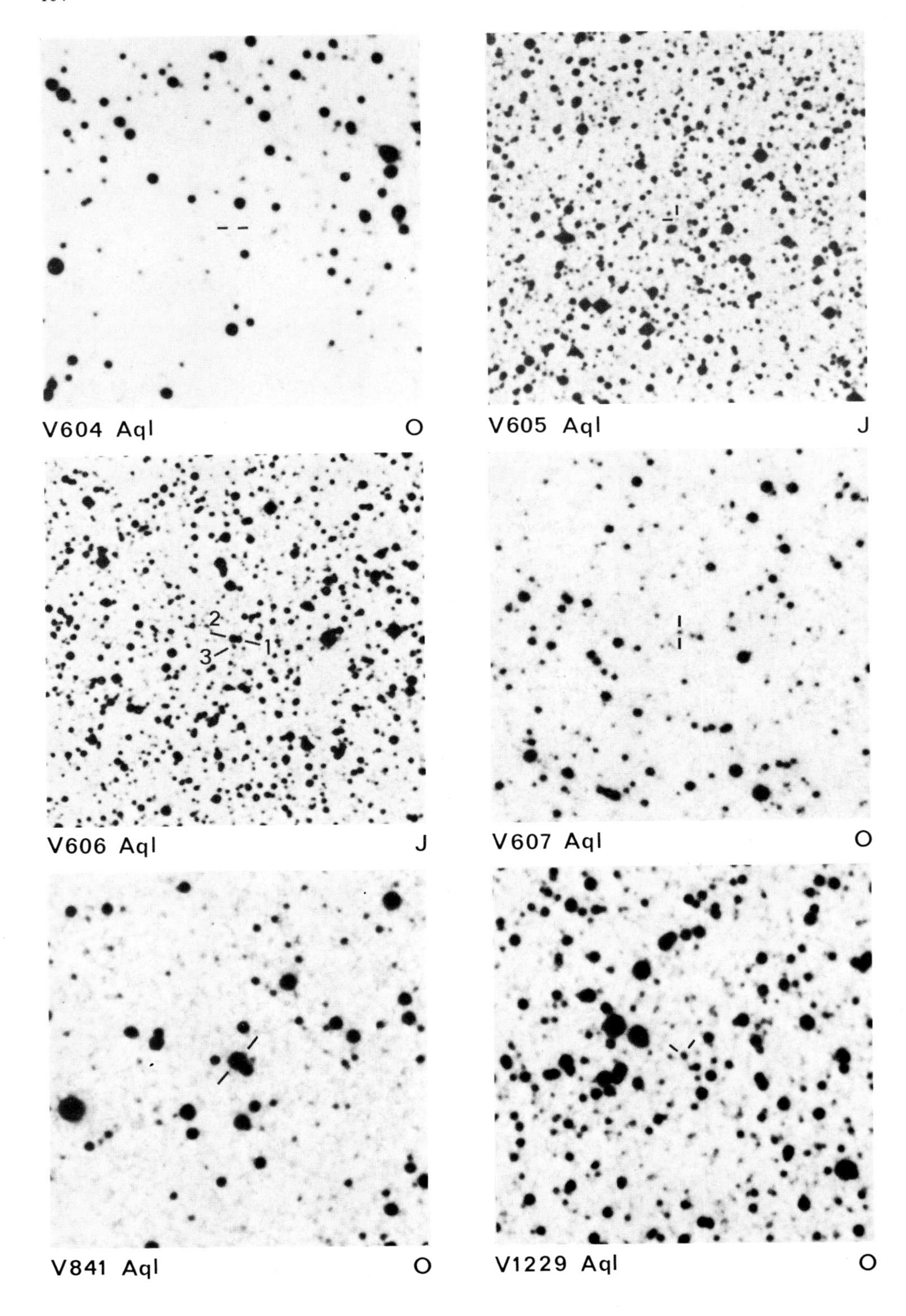

2
1
3
V604 Aql
O
V605 Aql
J
V606 Aql
J
V607 Aql
O
V841 Aql
O
V1229 Aql
O

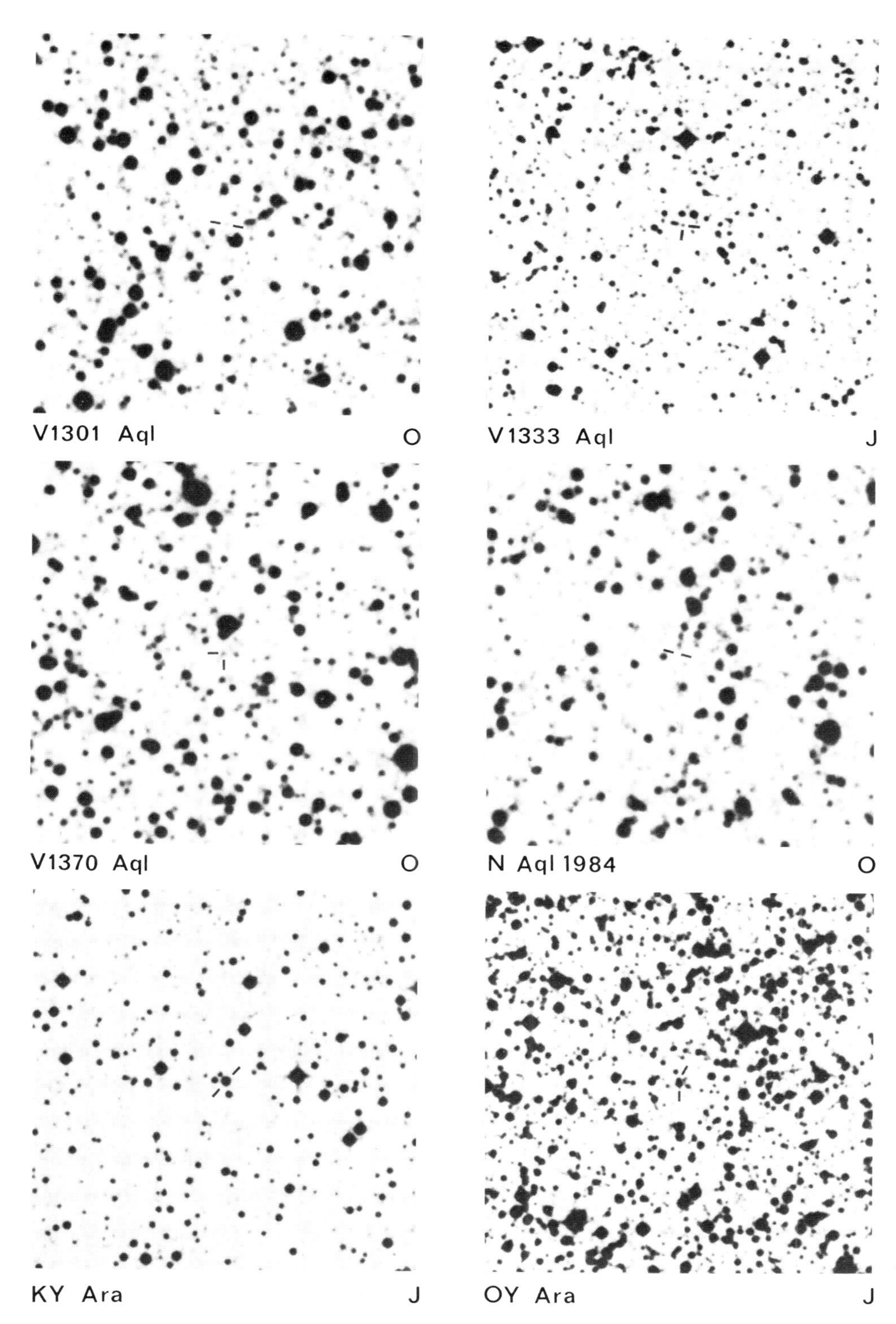
V1301 Aql
O
V1333 Aql
J
V1370 Aql
O
N Aql 1984
O
KY Ara
J
OY Ara
J

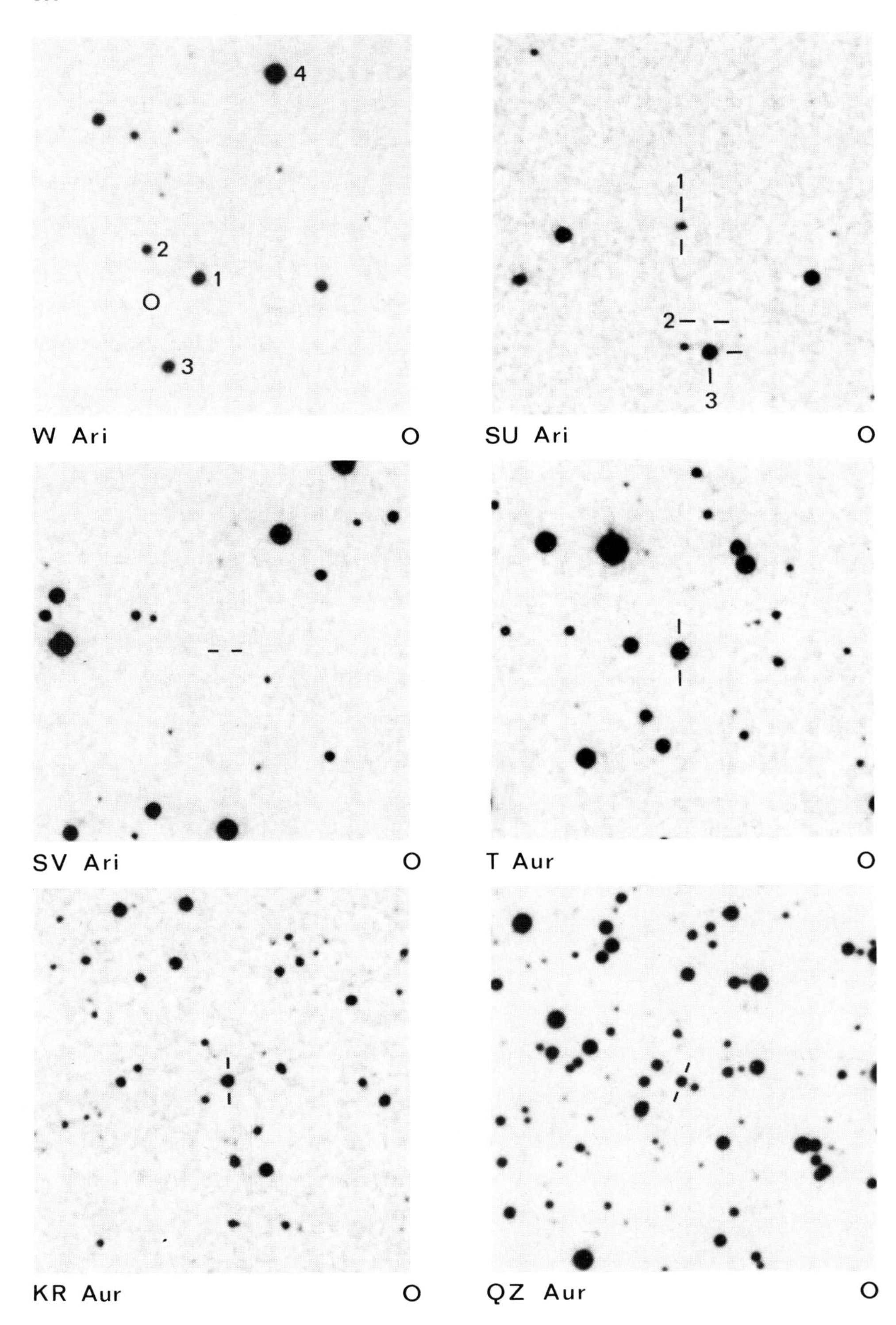
4
2
1
O
3
W Ari
O
1
2
3
SU Ari
O
SV Ari
O
T Aur
O
KR Aur
O
QZ Aur
O

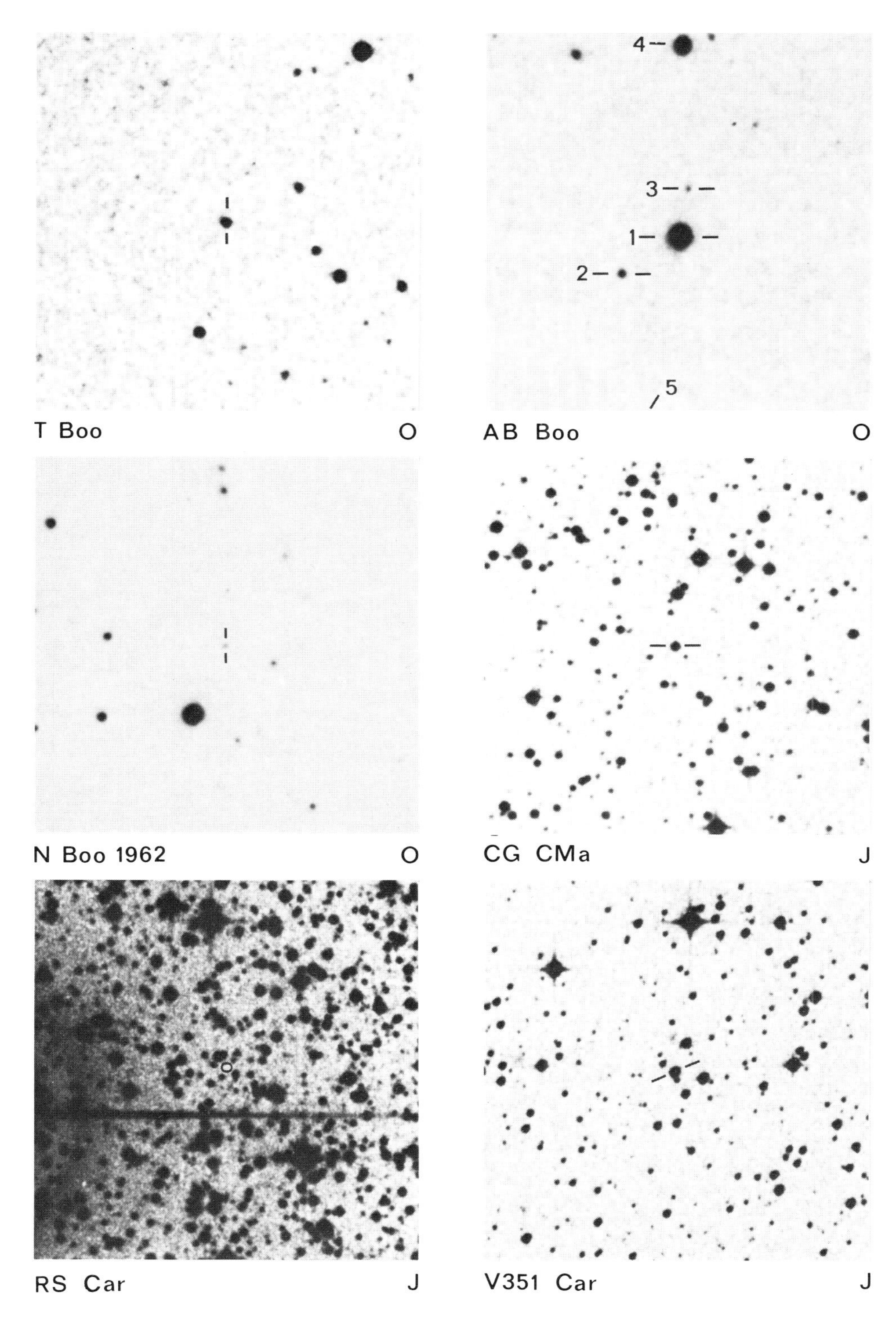

T Boo O

AB Boo O

N Boo 1962 O

CG CMa J

RS Car J

V351 Car J

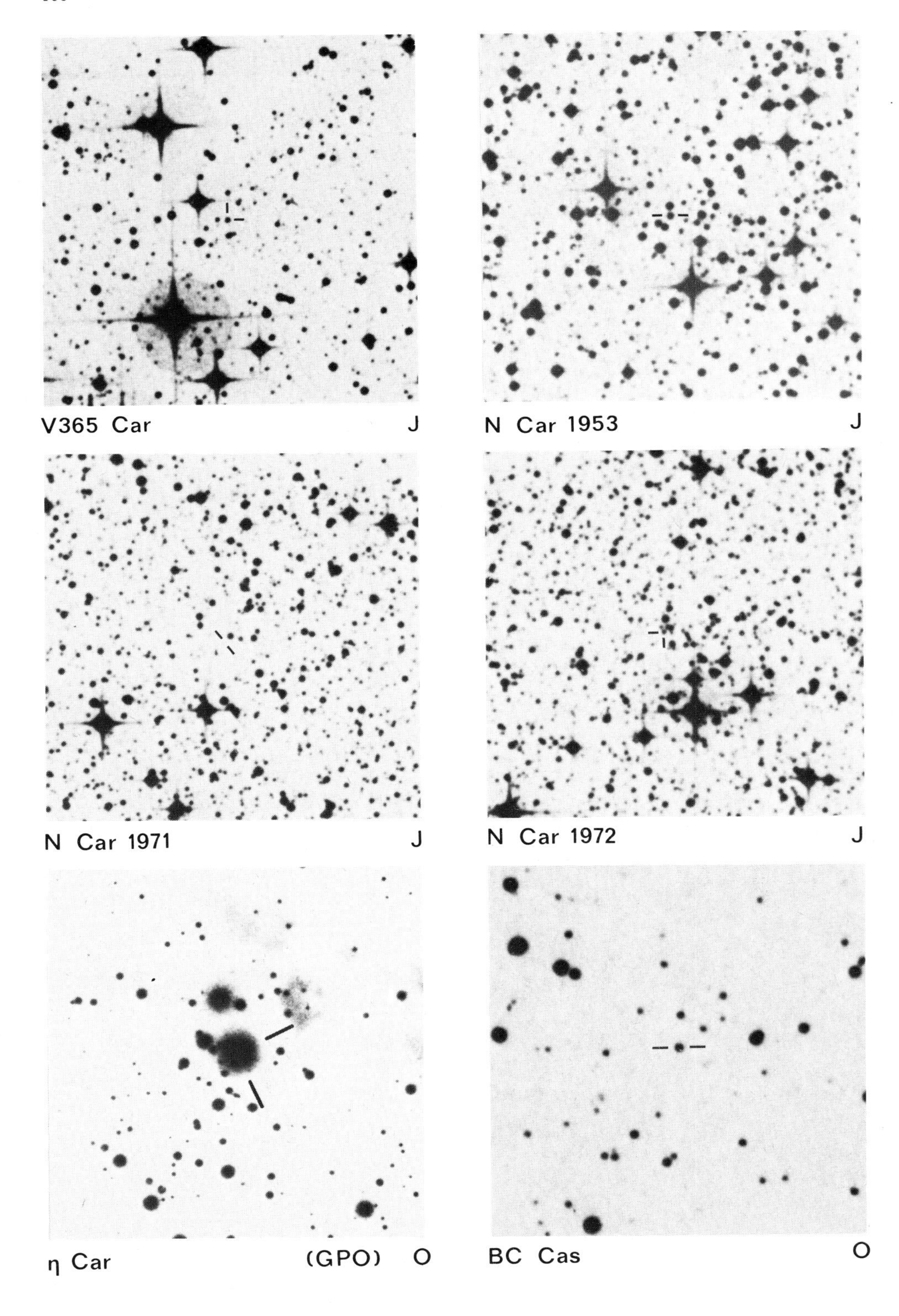
V365 Car
J
N Car 1953
J
N Car 1971
J
N Car 1972
J
η Car
(GPO)
O
BC Cas
O

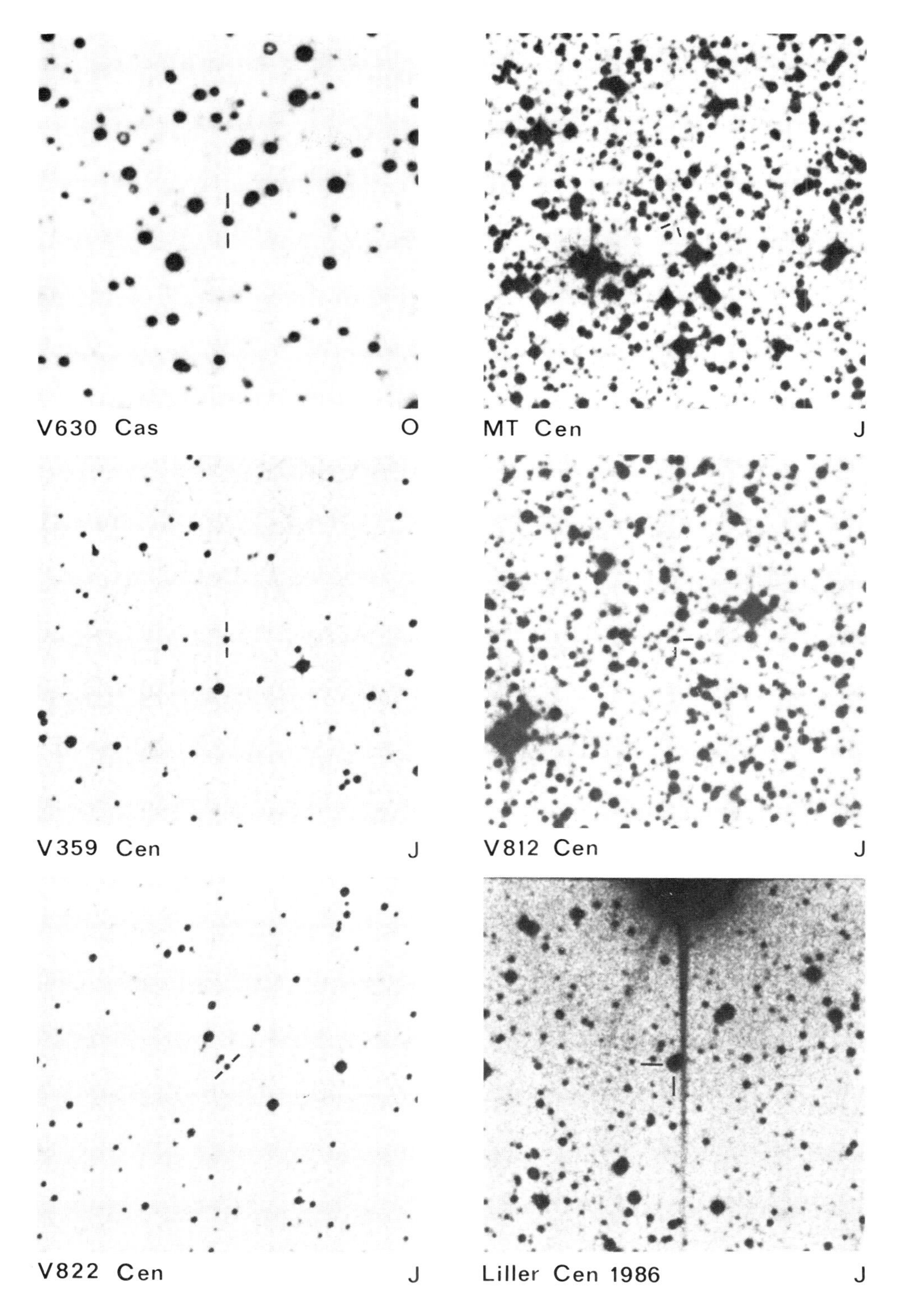
V630 Cas O
MT Cen J
V359 Cen J
V812 Cen J
V822 Cen J
Liller Cen 1986 J

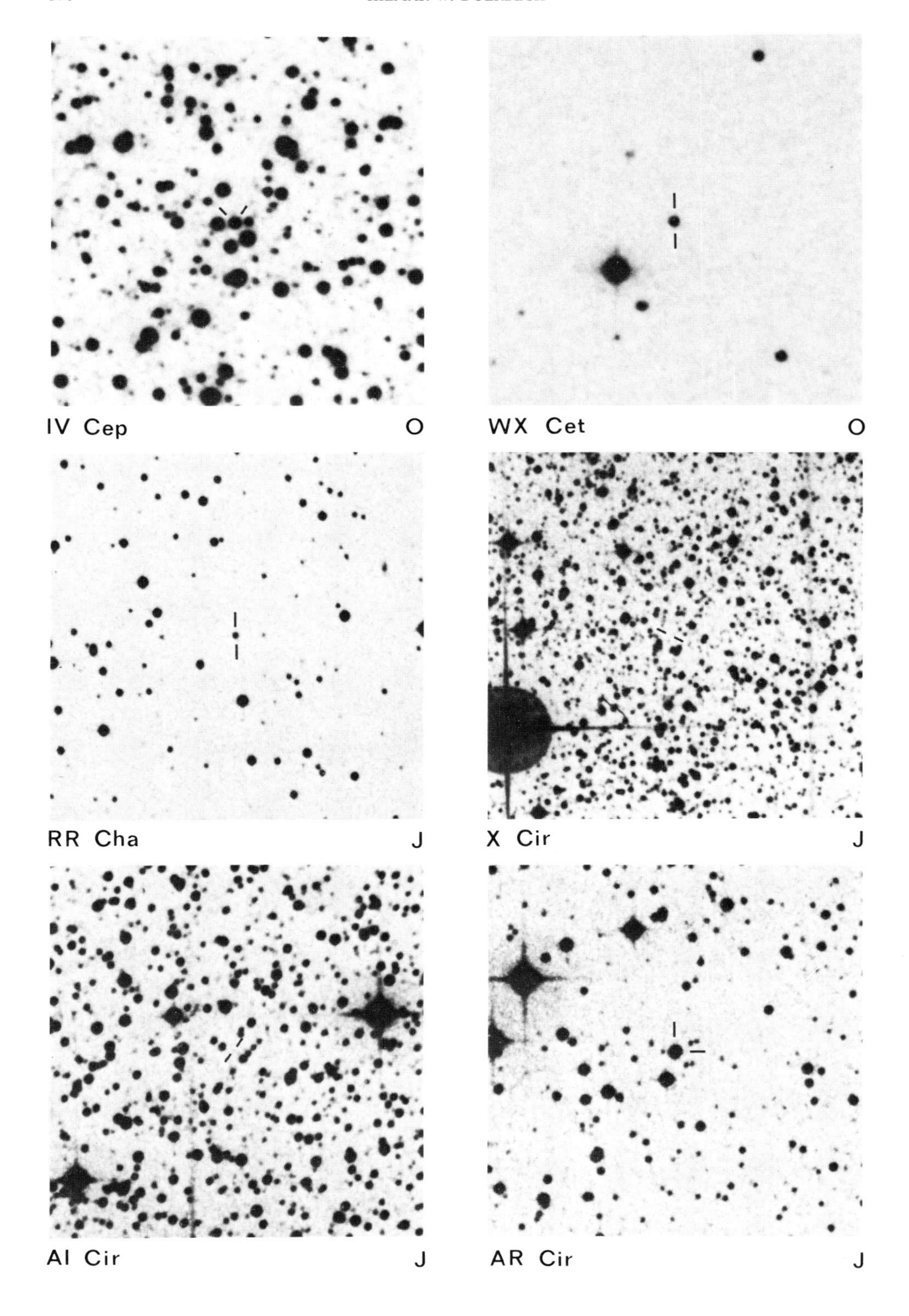
IV Cep
O
WX Cet
O
RR Cha
J
X Cir
J
AI Cir
J
AR Cir
J

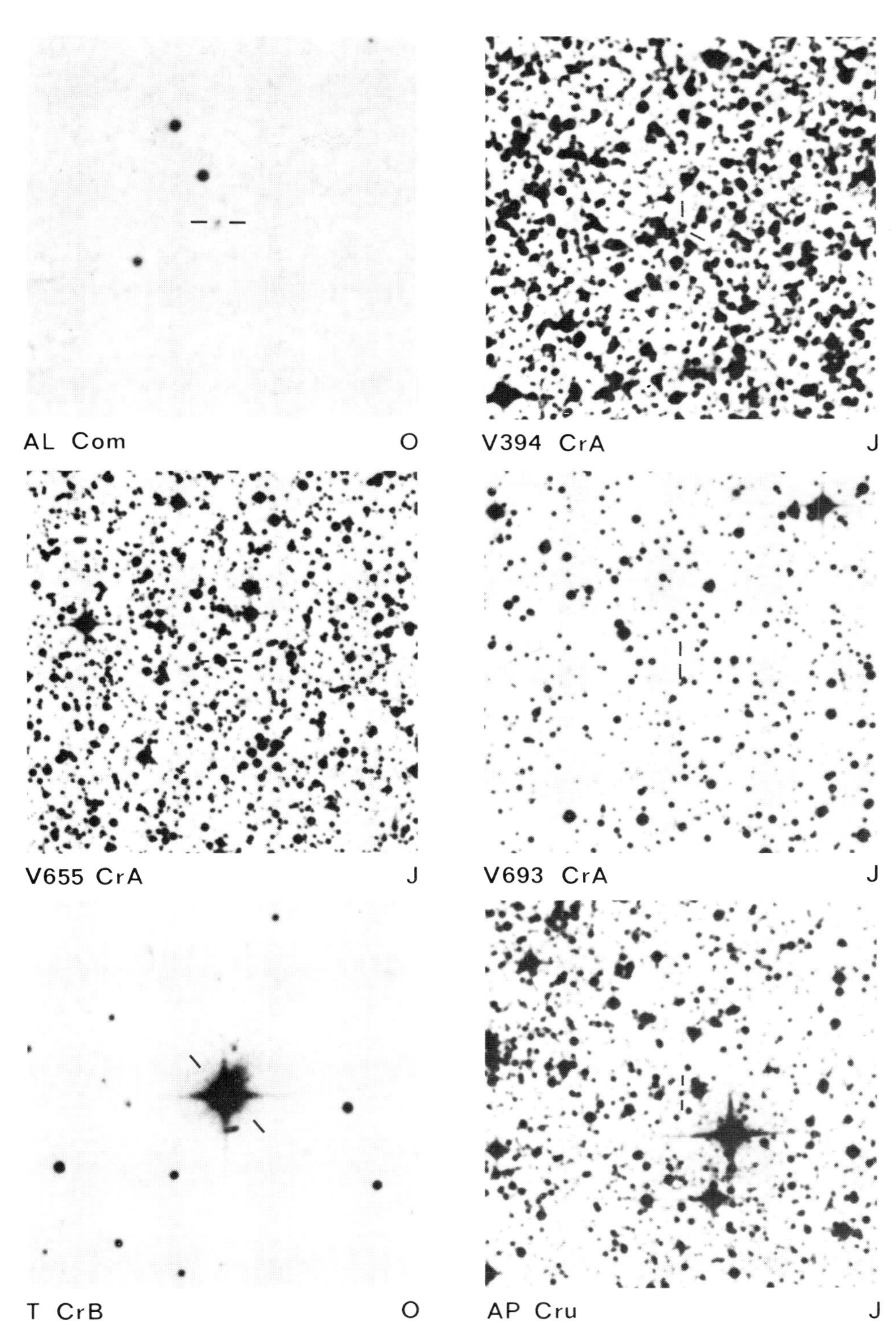

AL Com O

V394 CrA J

V655 CrA J

V693 CrA J

T CrB O

AP Cru J

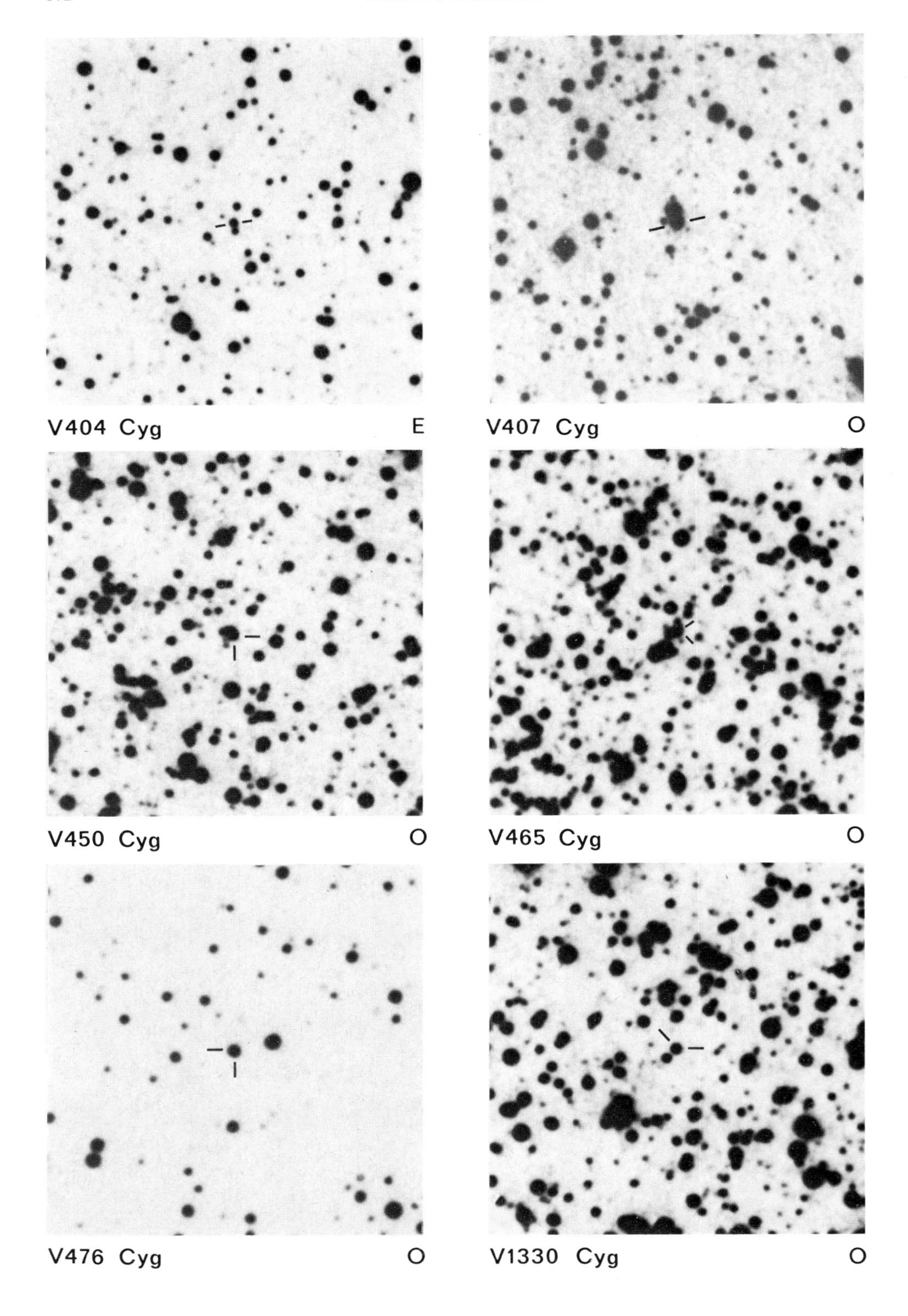
V404 Cyg
E
V407 Cyg
O
V450 Cyg
O
V465 Cyg
O
V476 Cyg
O
V1330 Cyg
O

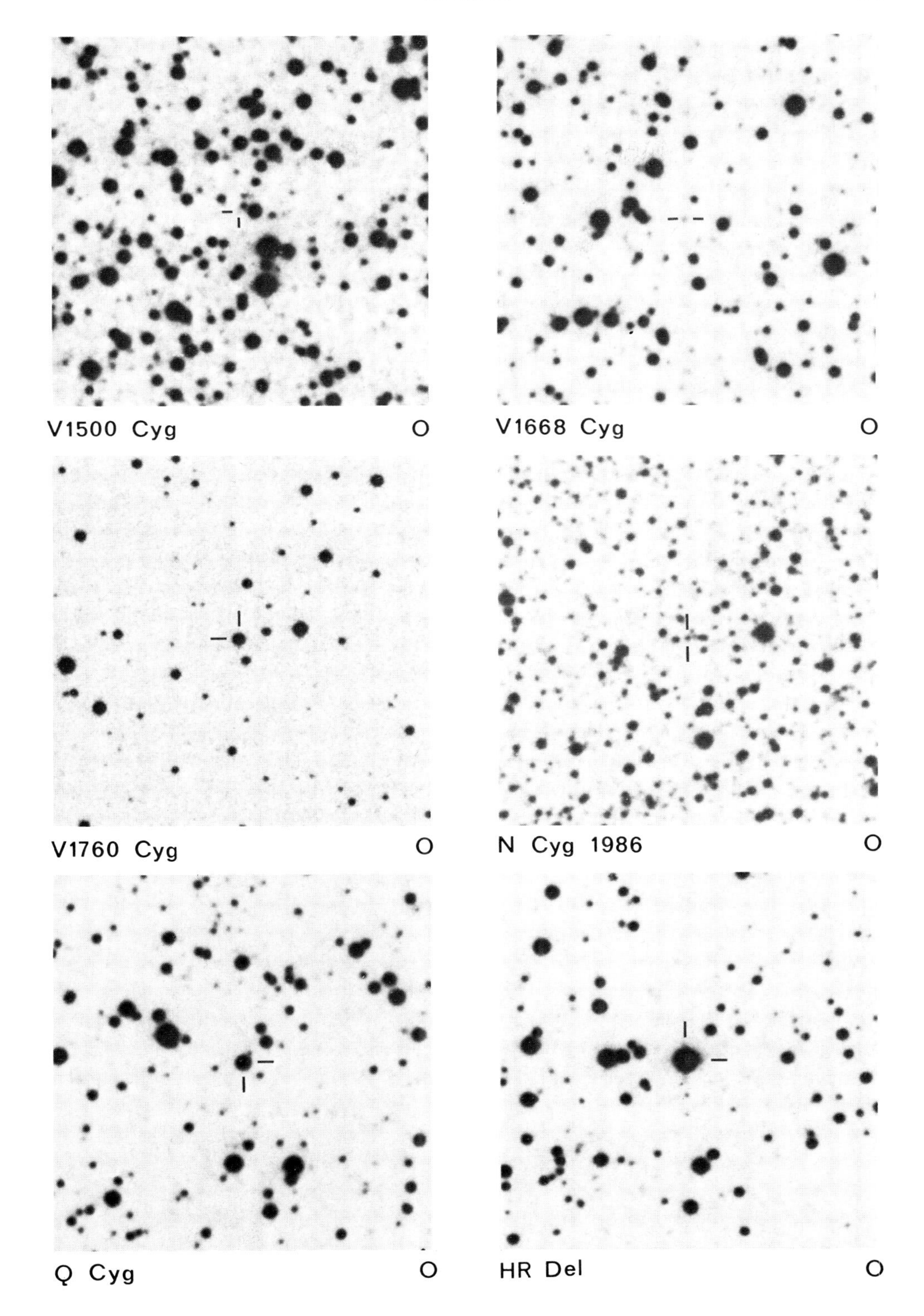
V1500 Cyg
O
V1668 Cyg
O
V1760 Cyg
O
N Cyg 1986
O
Q Cyg
O
HR Del
O

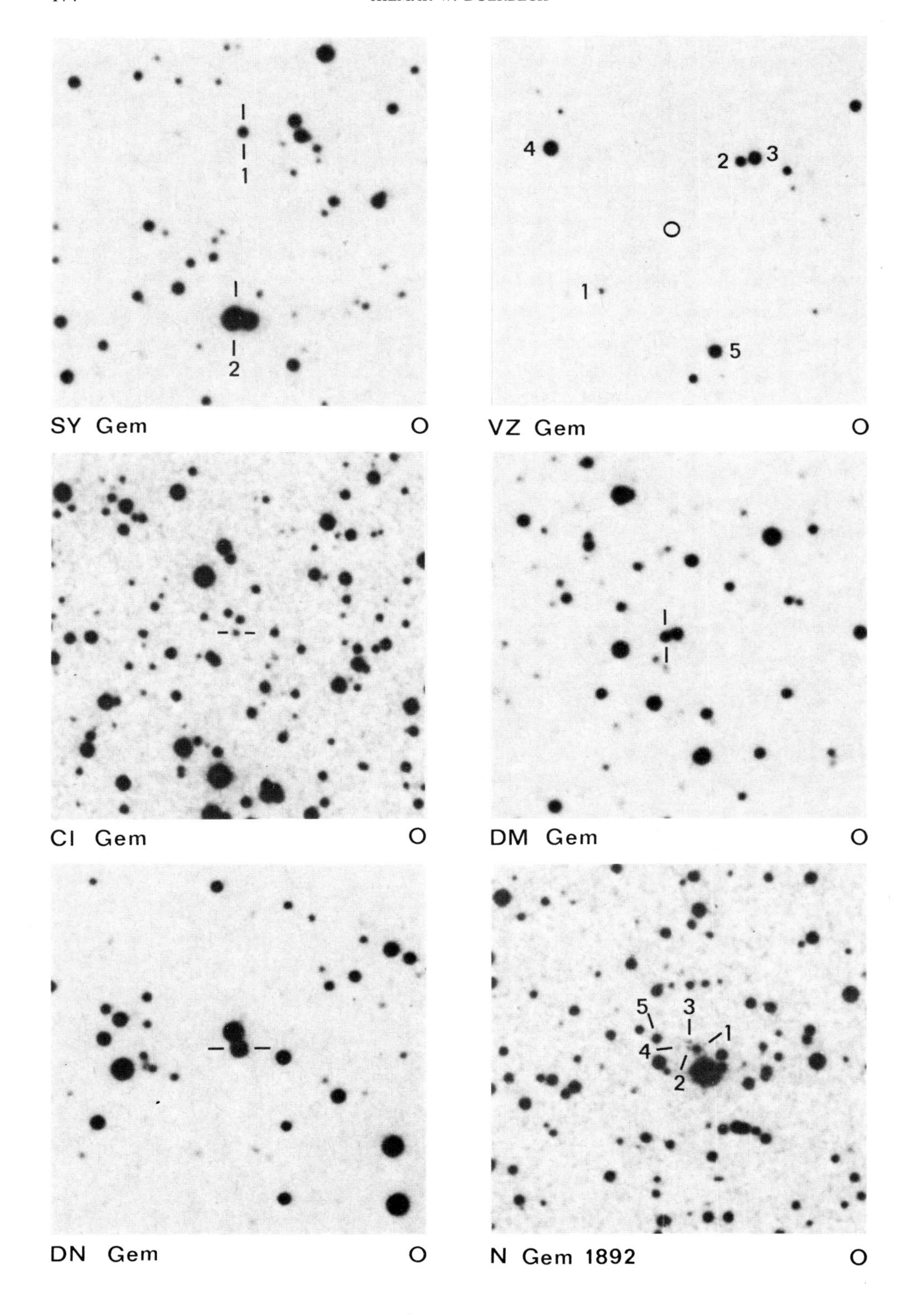
1
2
SY Gem
O
4
2
3
1
5
VZ Gem
O
CI Gem
O
DM Gem
O
DN Gem
O
5
3
1
4
2
N Gem 1892
O

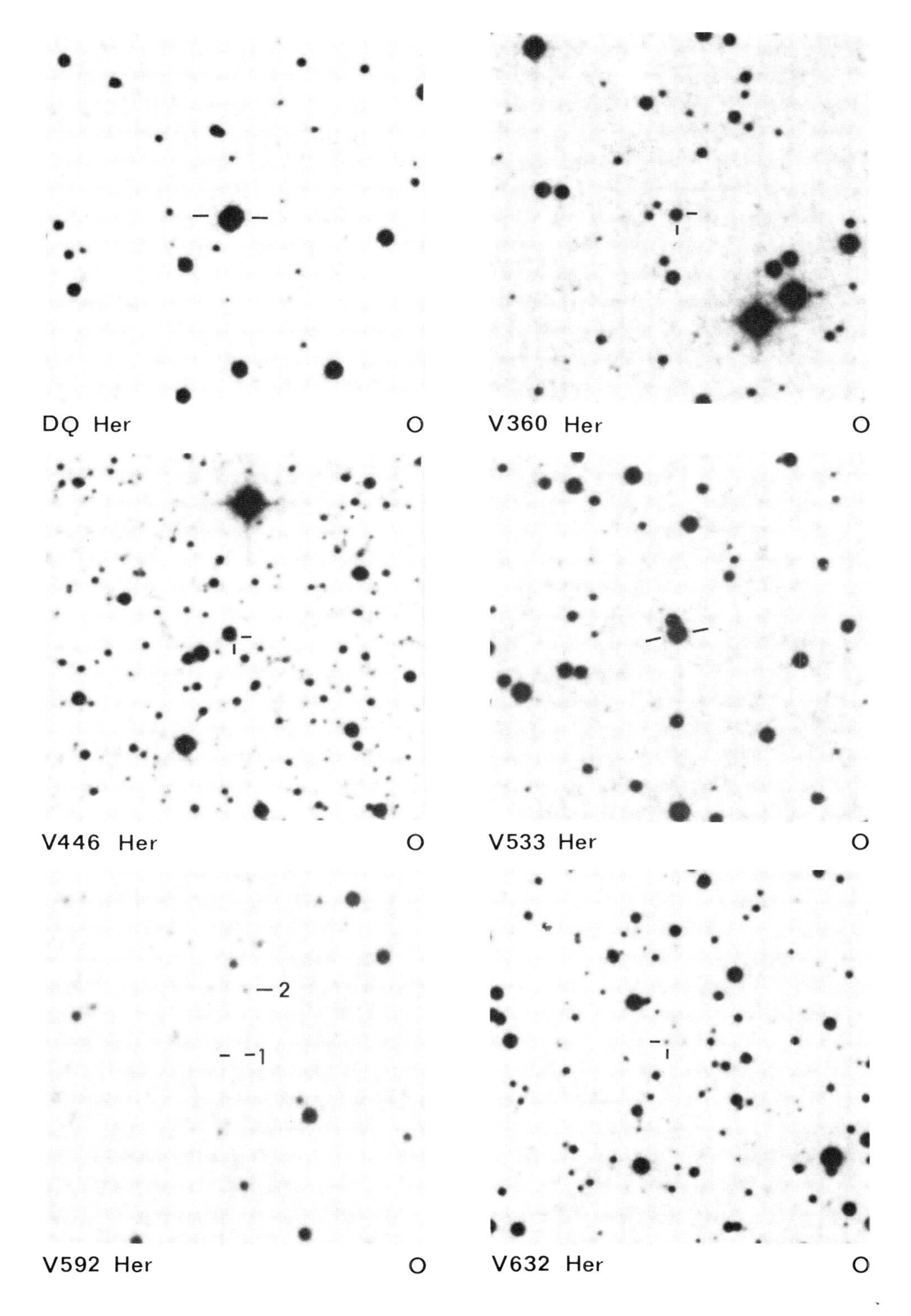
DQ Her
O
V360 Her
O
V446 Her
O
V533 Her
O
2
1
V592 Her
O
V632 Her
O

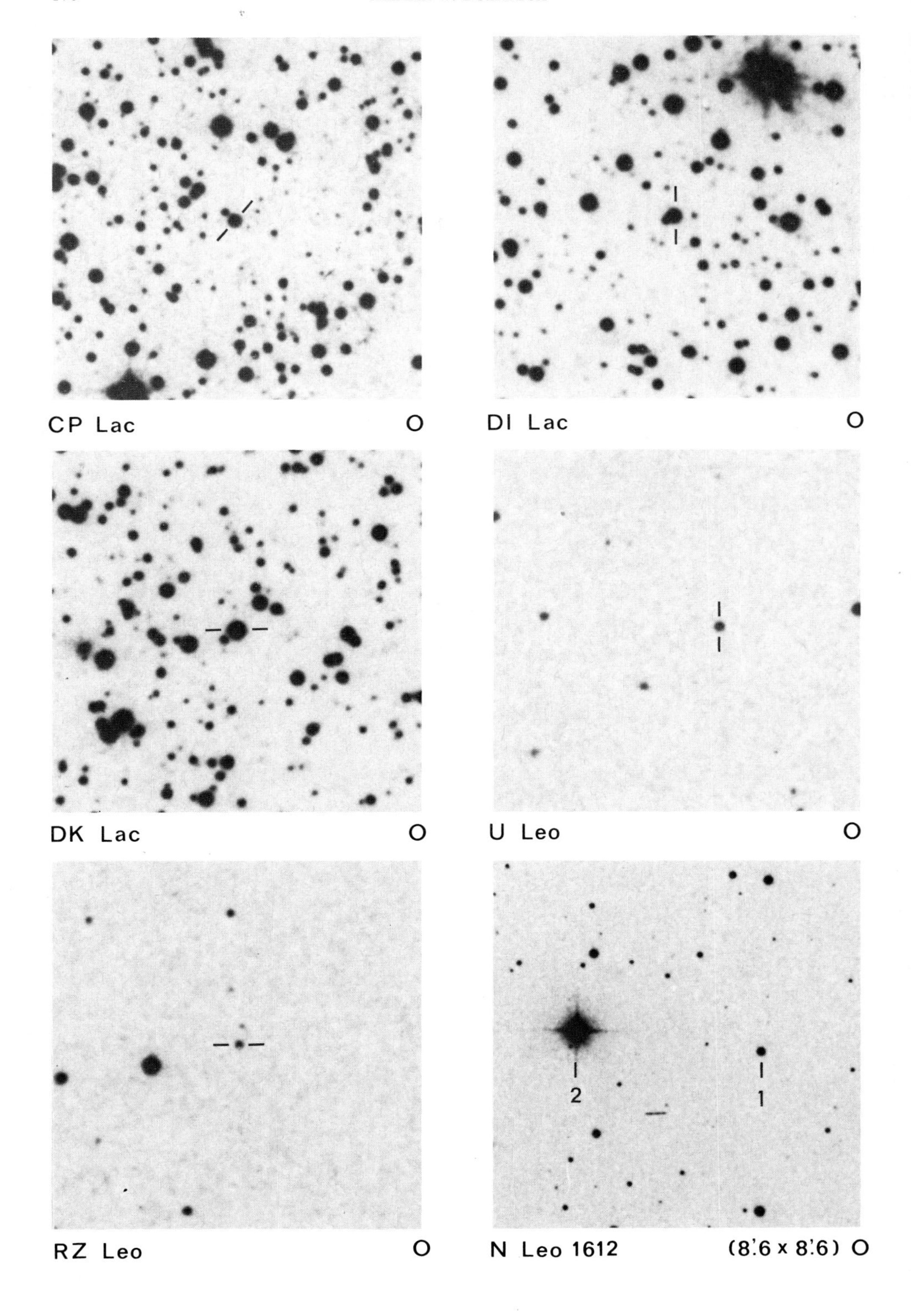
CP Lac
O
DI Lac
O
DK Lac
O
U Leo
O
RZ Leo
O
2
1
N Leo 1612
(8!6 x 8!6) O

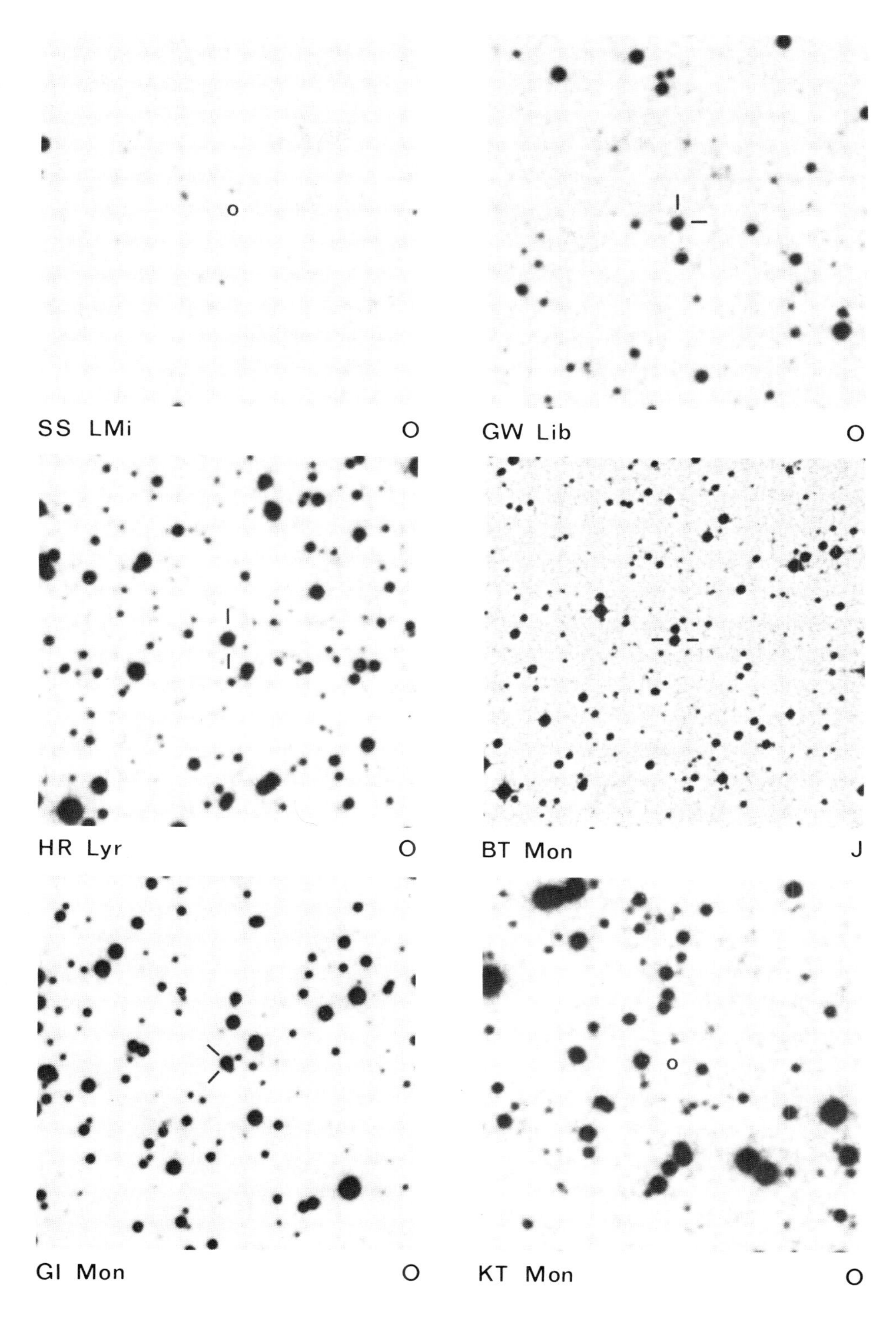
o
SS LMi
O
GW Lib
O
HR Lyr
O
BT Mon
J
GI Mon
O
o
KT Mon
O

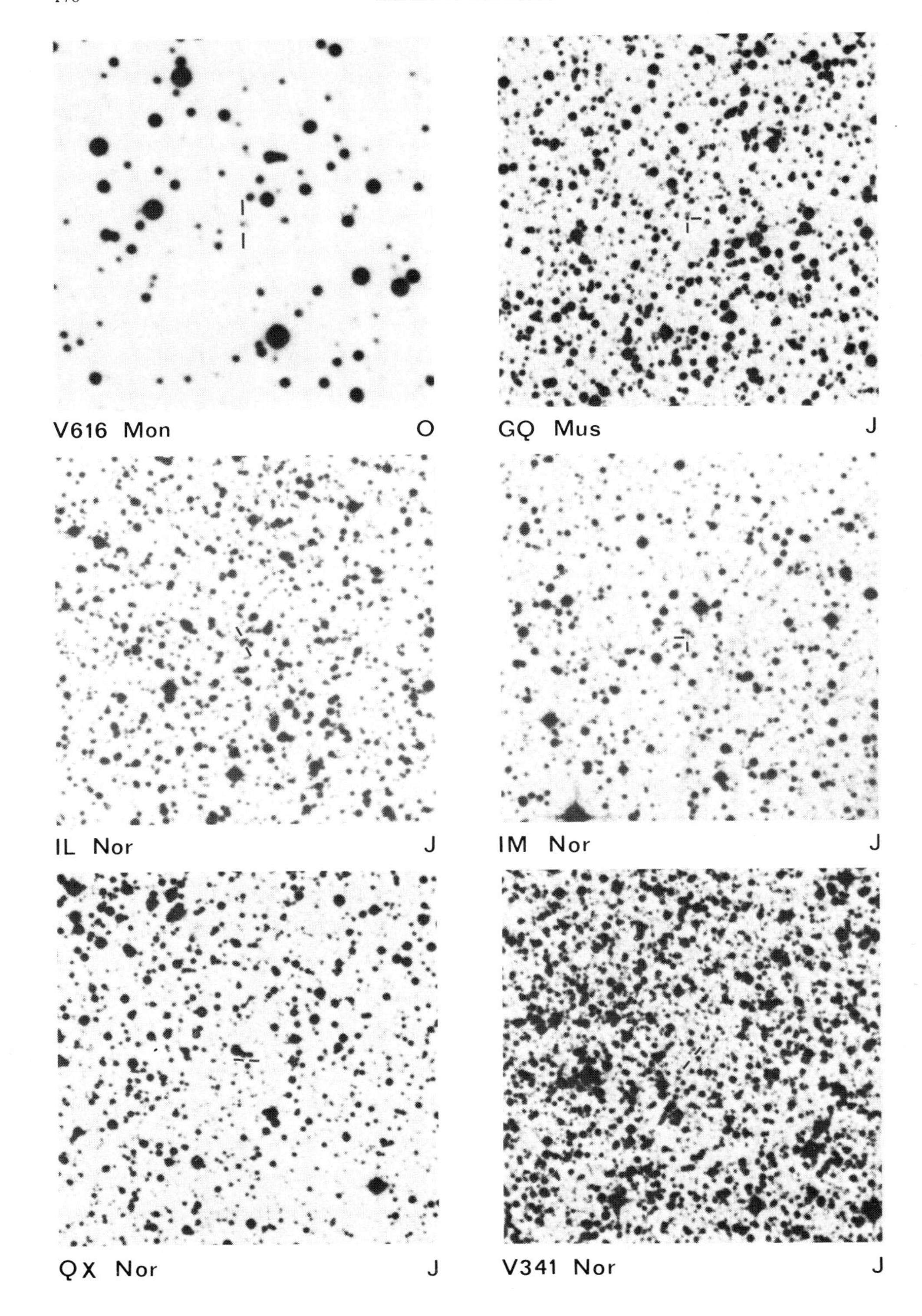
V616 Mon
O
GQ Mus
J
IL Nor
J
IM Nor
J
QX Nor
J
V341 Nor
J

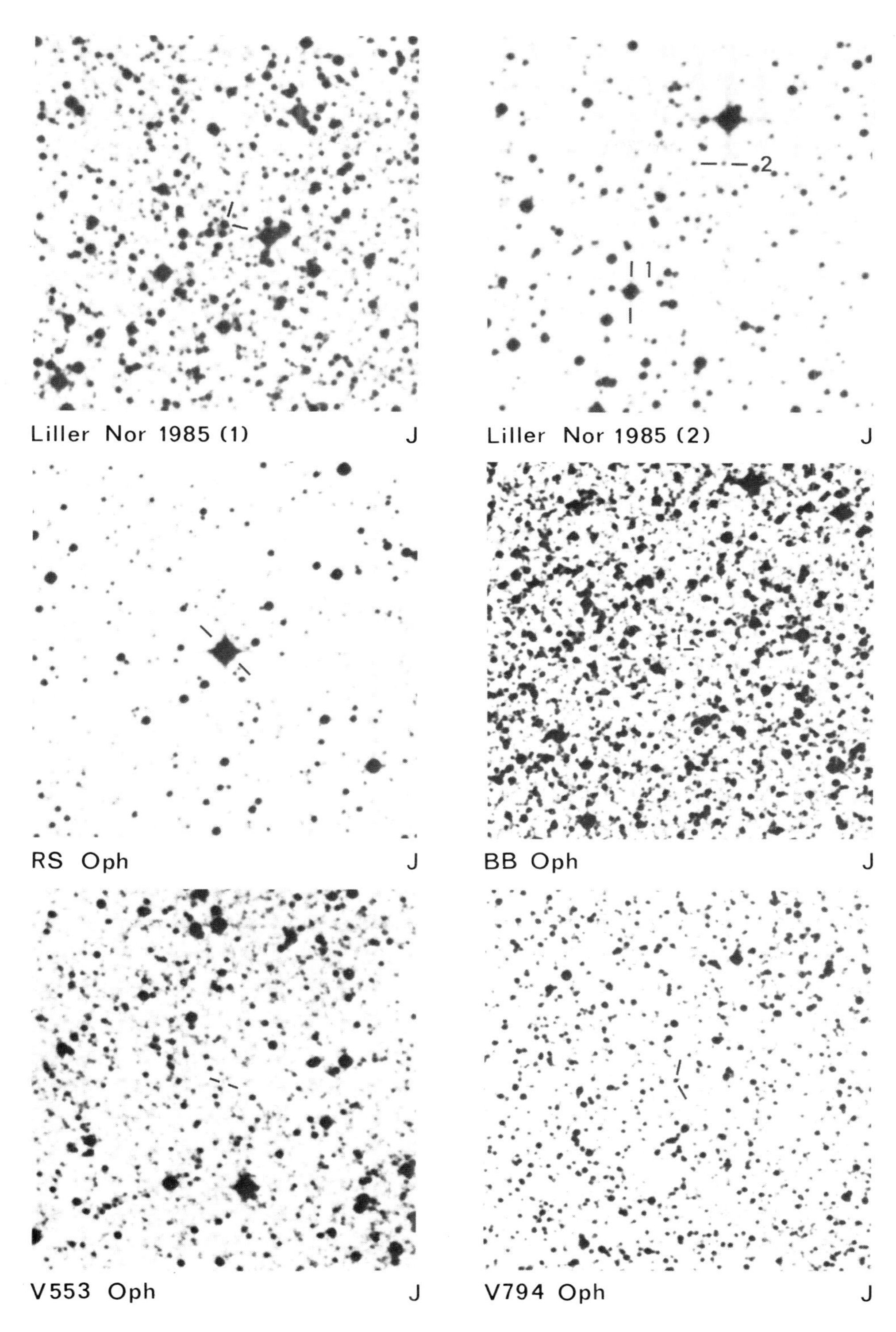

Liller Nor 1985 (1) J

Liller Nor 1985 (2) J

RS Oph J

BB Oph J

V553 Oph J

V794 Oph J

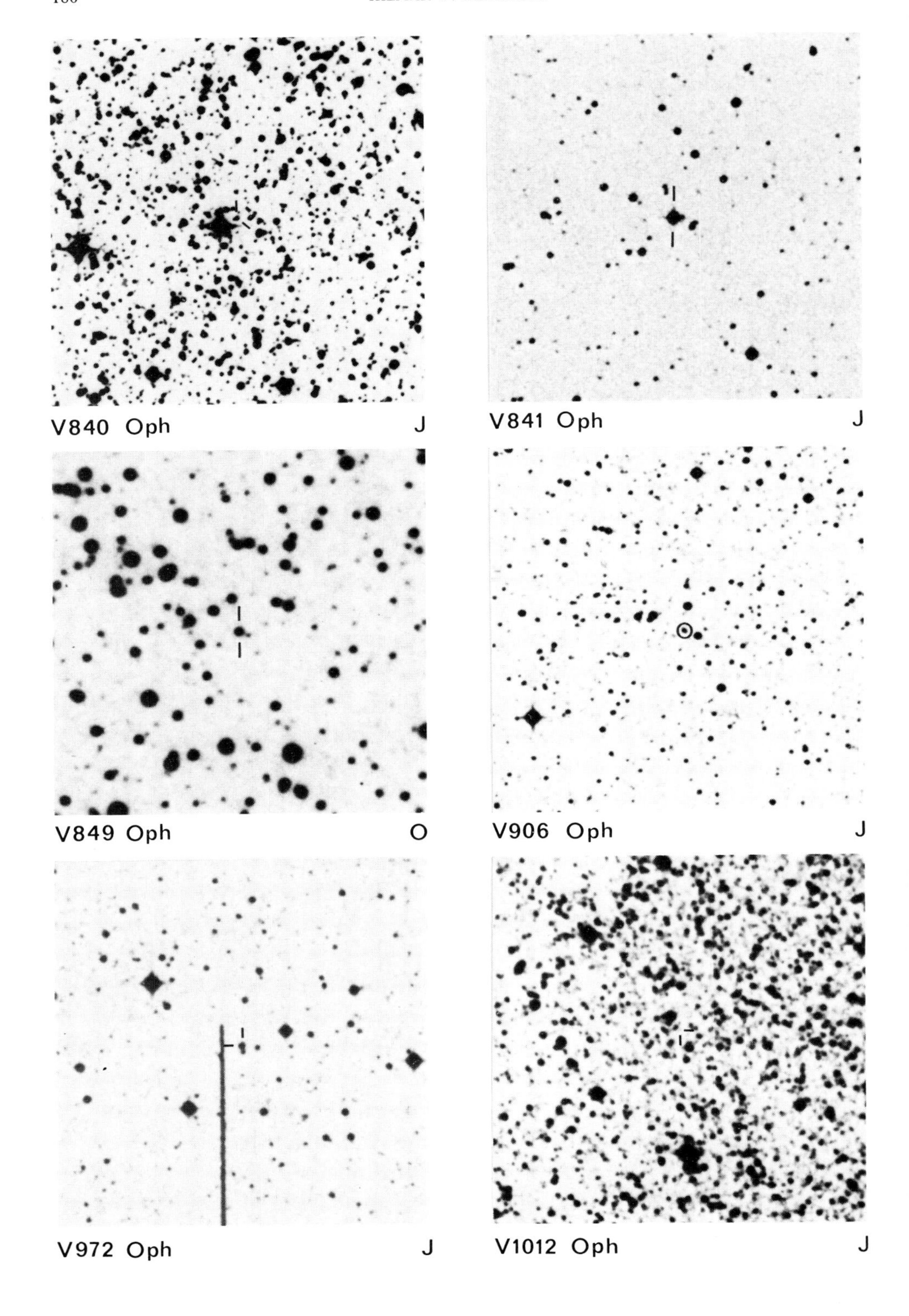
V840 Oph J
V841 Oph J
V849 Oph O
V906 Oph J
V972 Oph J
V1012 Oph J

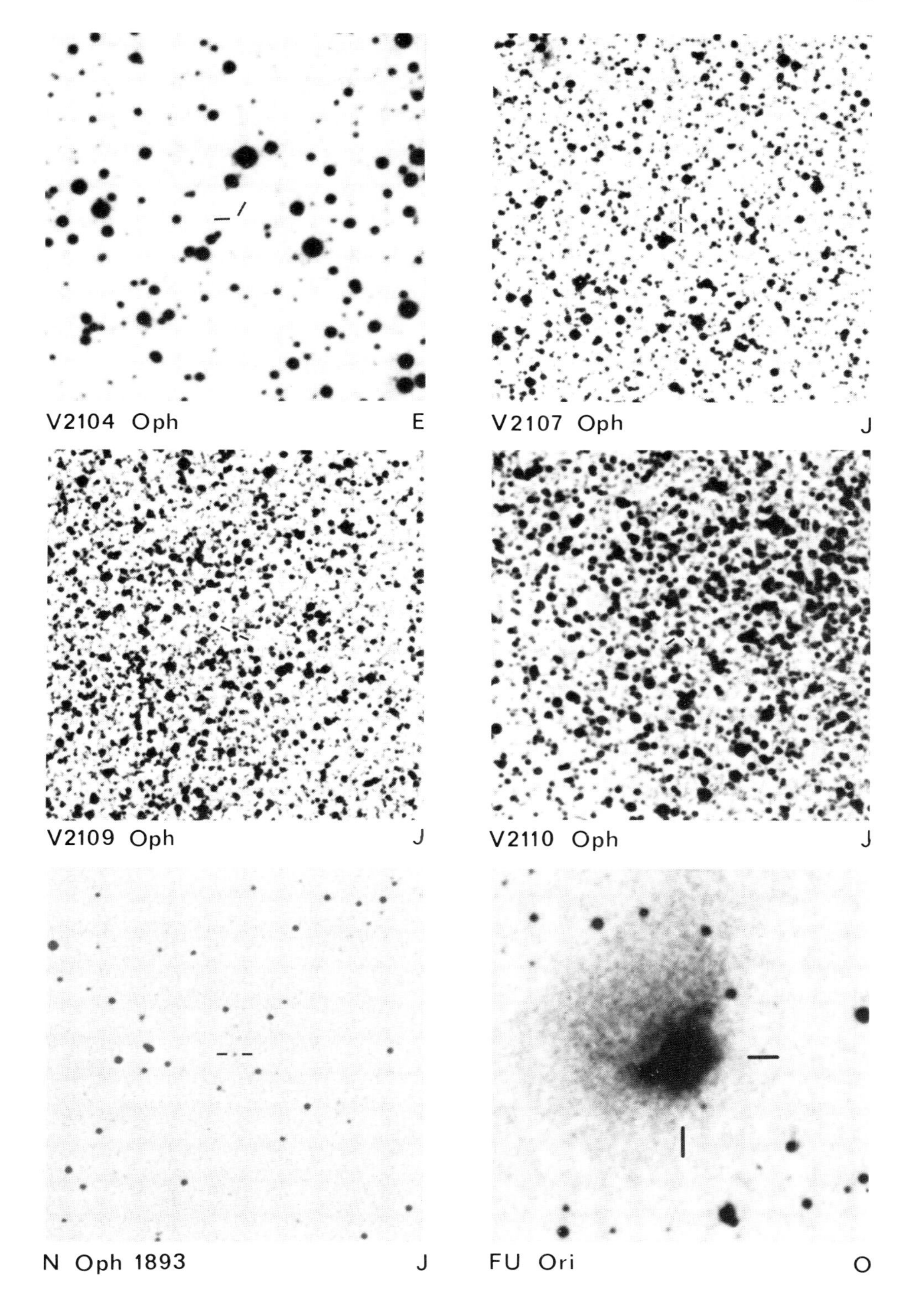

V2104 Oph E

V2107 Oph J

V2109 Oph J

V2110 Oph J

N Oph 1893 J

FU Ori O

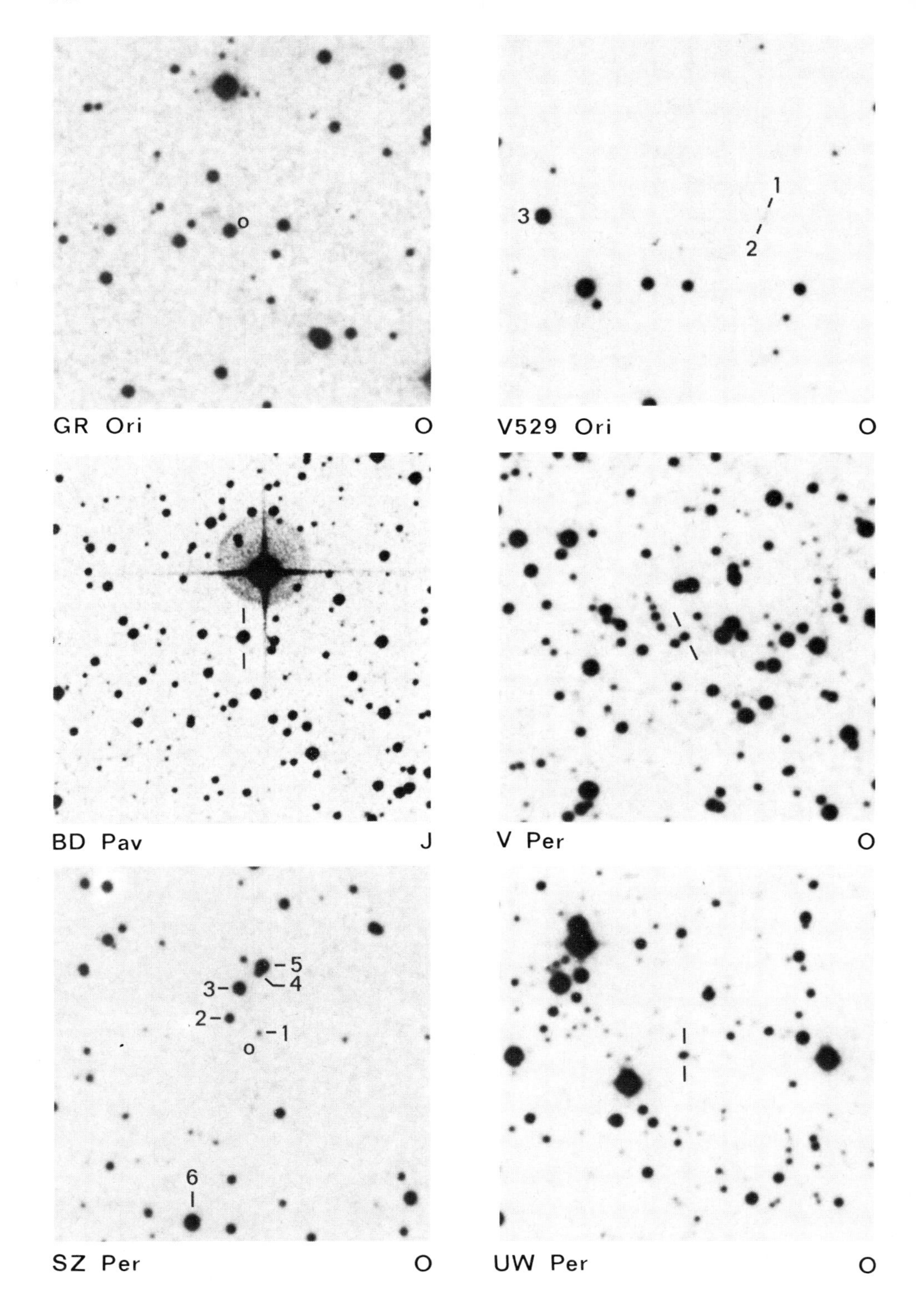

o
GR Ori
O
3
1
2
V529 Ori
O
BD Pav
J
V Per
O
5
4
3
2
1
o
6
SZ Per
O
UW Per
O

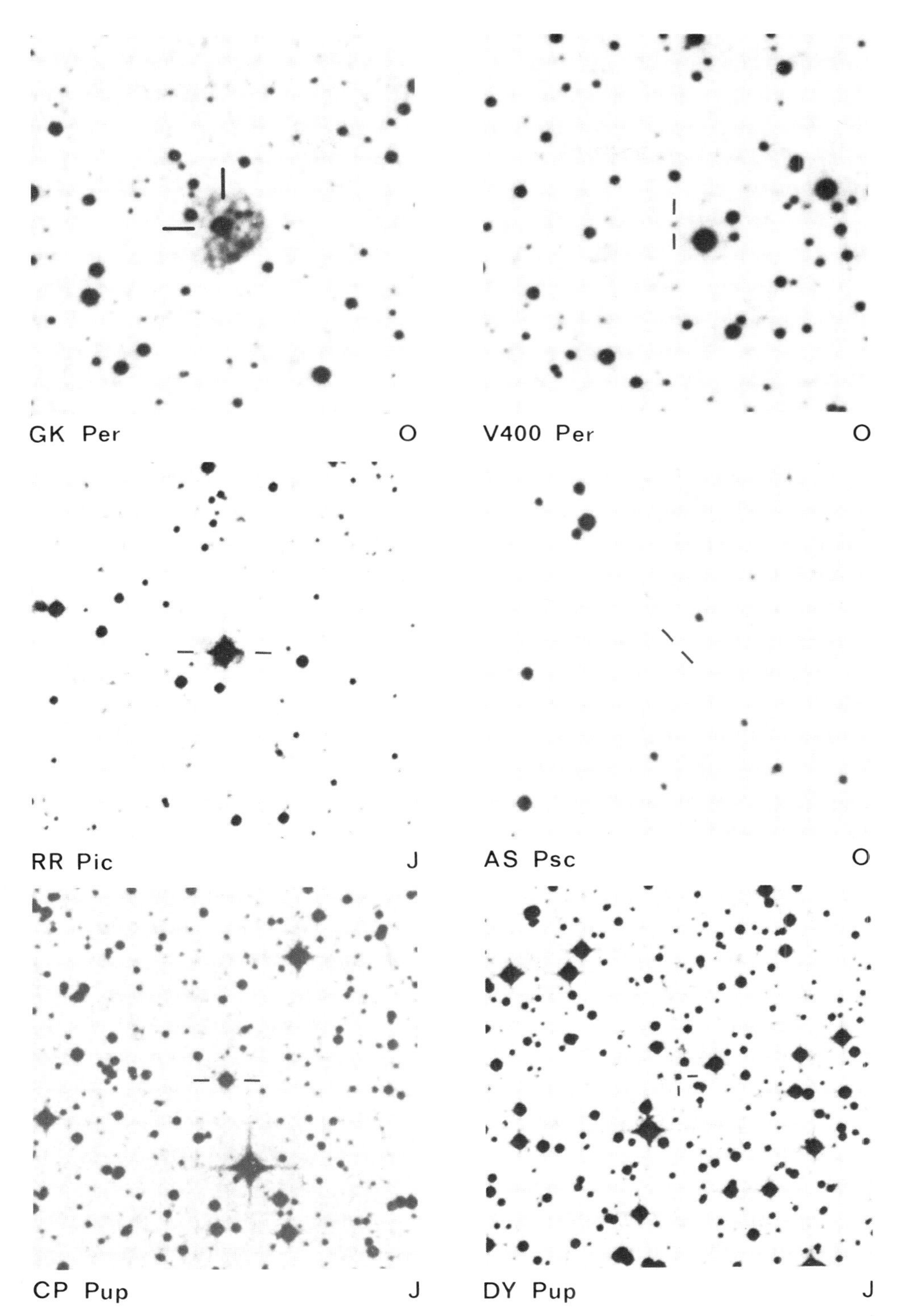
GK Per
O
V400 Per
O
RR Pic
J
AS Psc
O
CP Pup
J
DY Pup
J

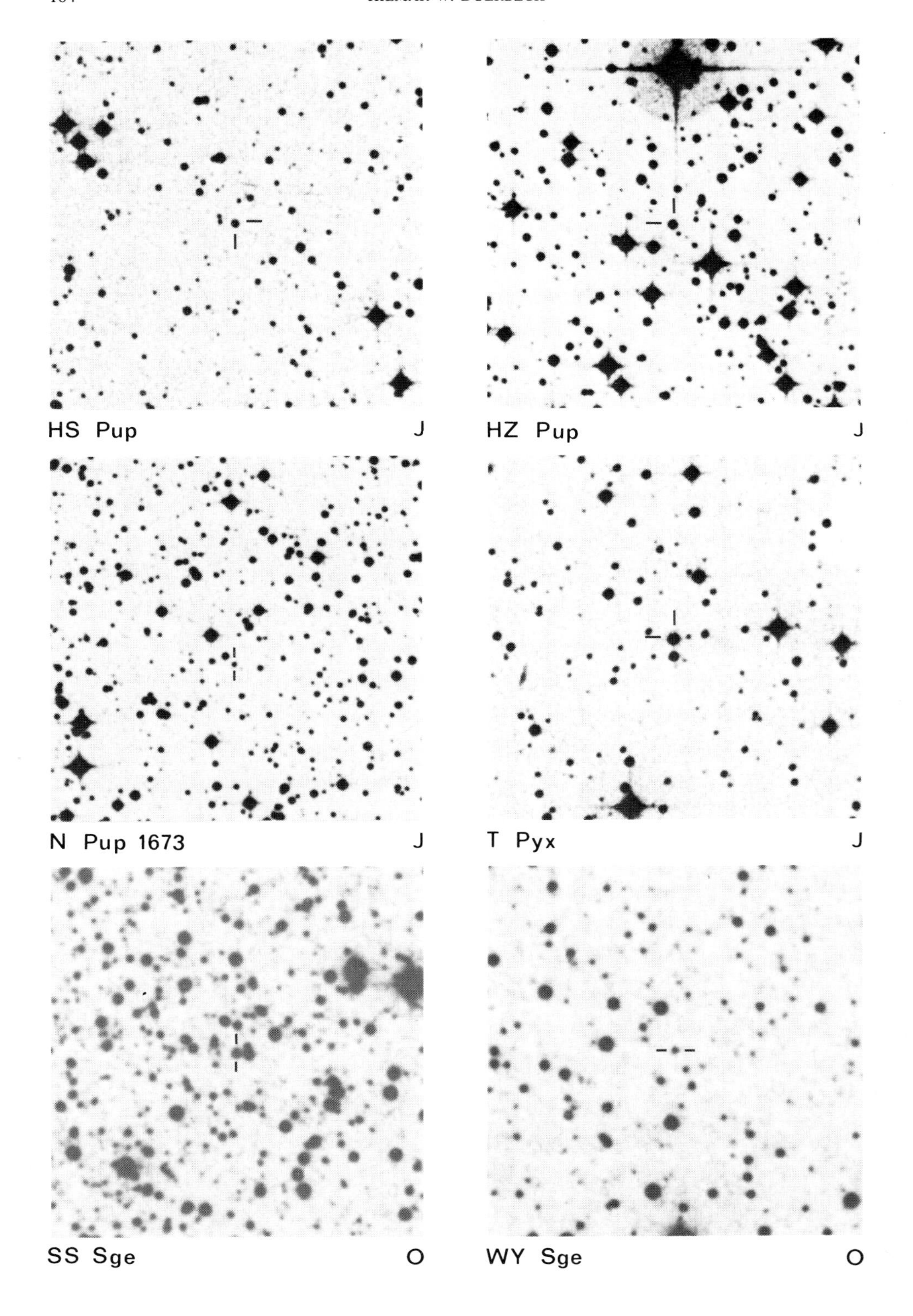
HS Pup J
HZ Pup J
N Pup 1673 J
T Pyx J
SS Sge O
WY Sge O

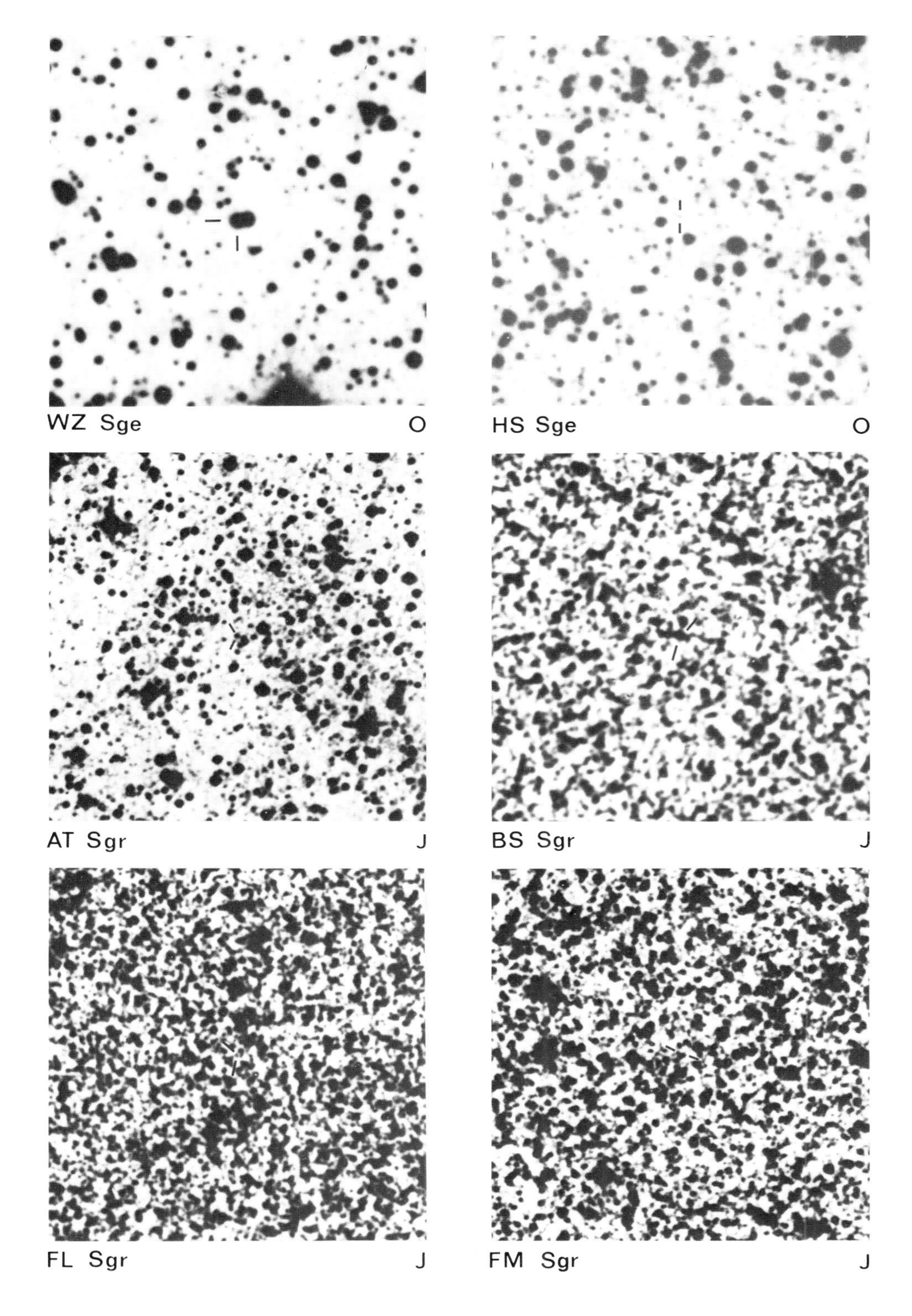
WZ Sge
O
HS Sge
O
AT Sgr
J
BS Sgr
J
FL Sgr
J
FM Sgr
J

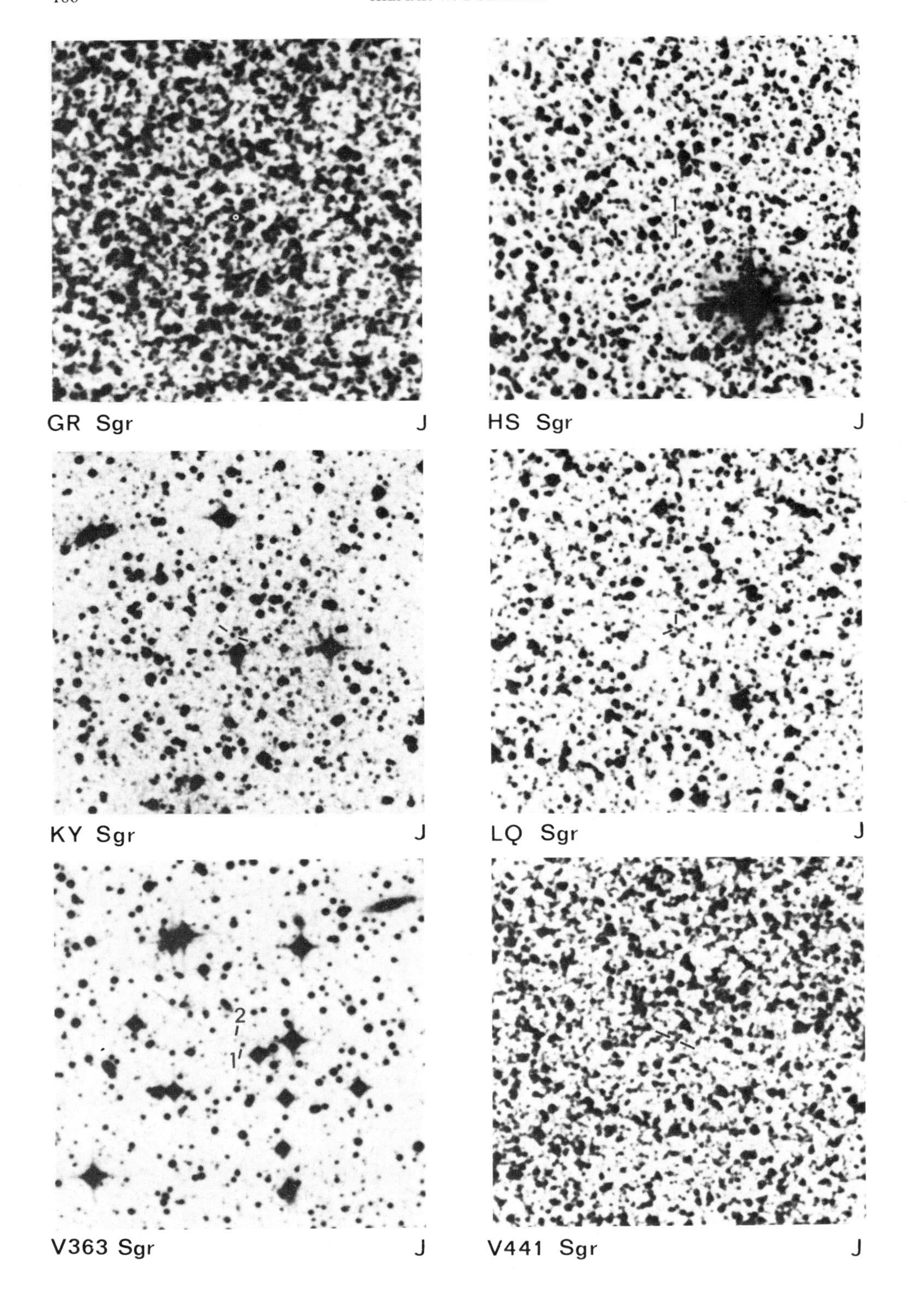
GR Sgr
J
HS Sgr
J
KY Sgr
J
LQ Sgr
J
2
1
V363 Sgr
J
V441 Sgr
J

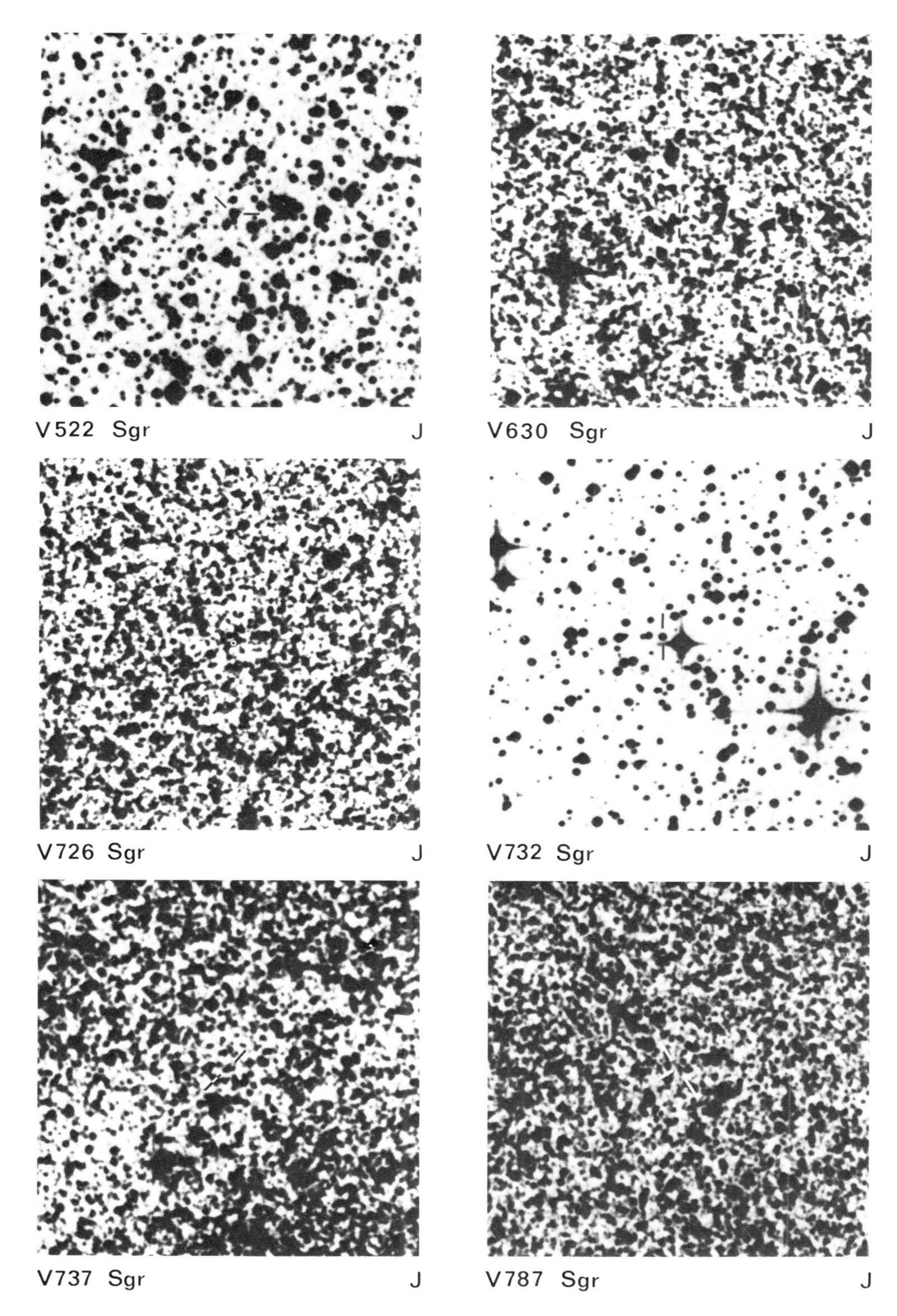
V522 Sgr J
V630 Sgr J
V726 Sgr J
V732 Sgr J
V737 Sgr J
V787 Sgr J

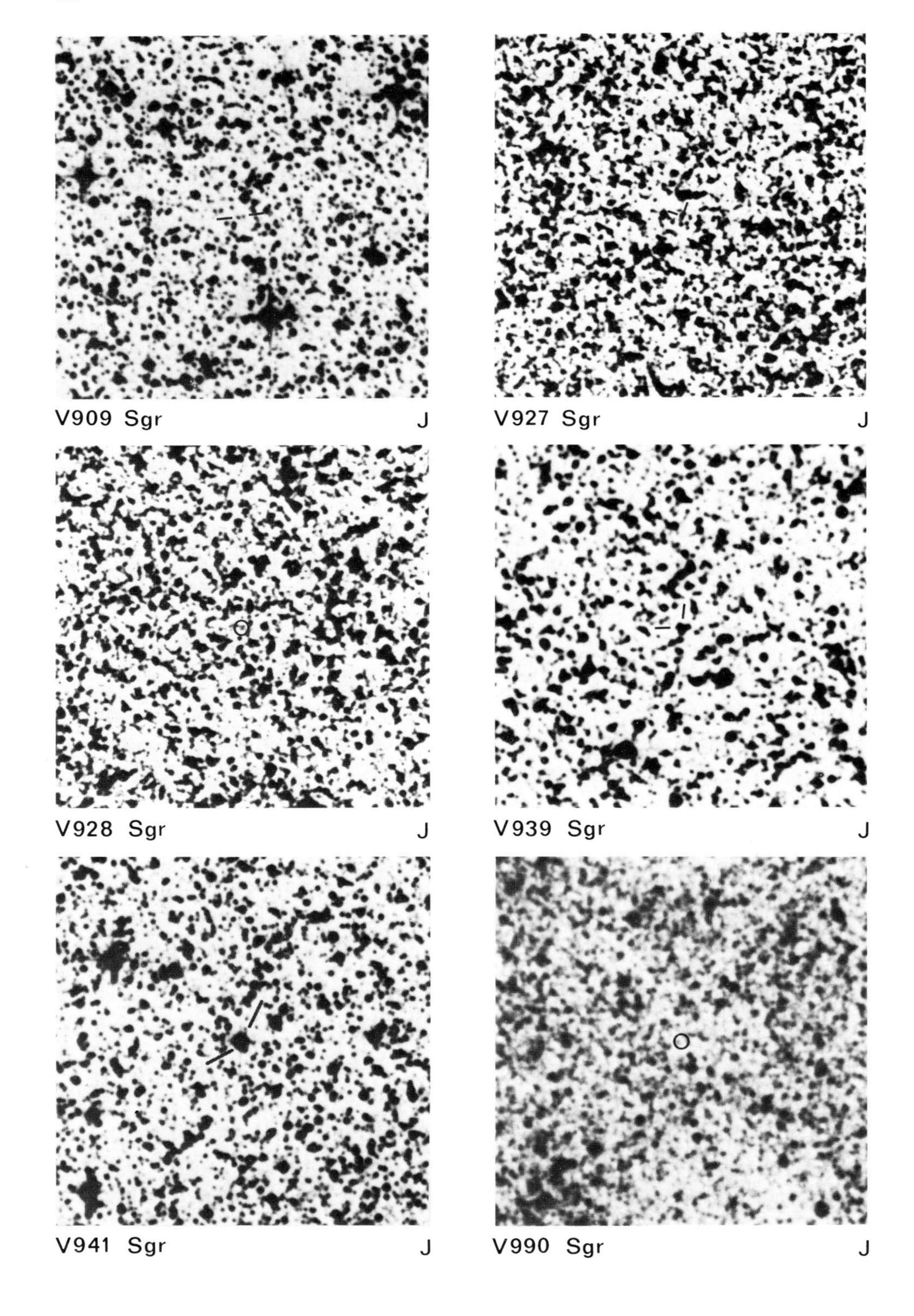
V909 Sgr J
V927 Sgr J
V928 Sgr J
V939 Sgr J
V941 Sgr J
V990 Sgr J

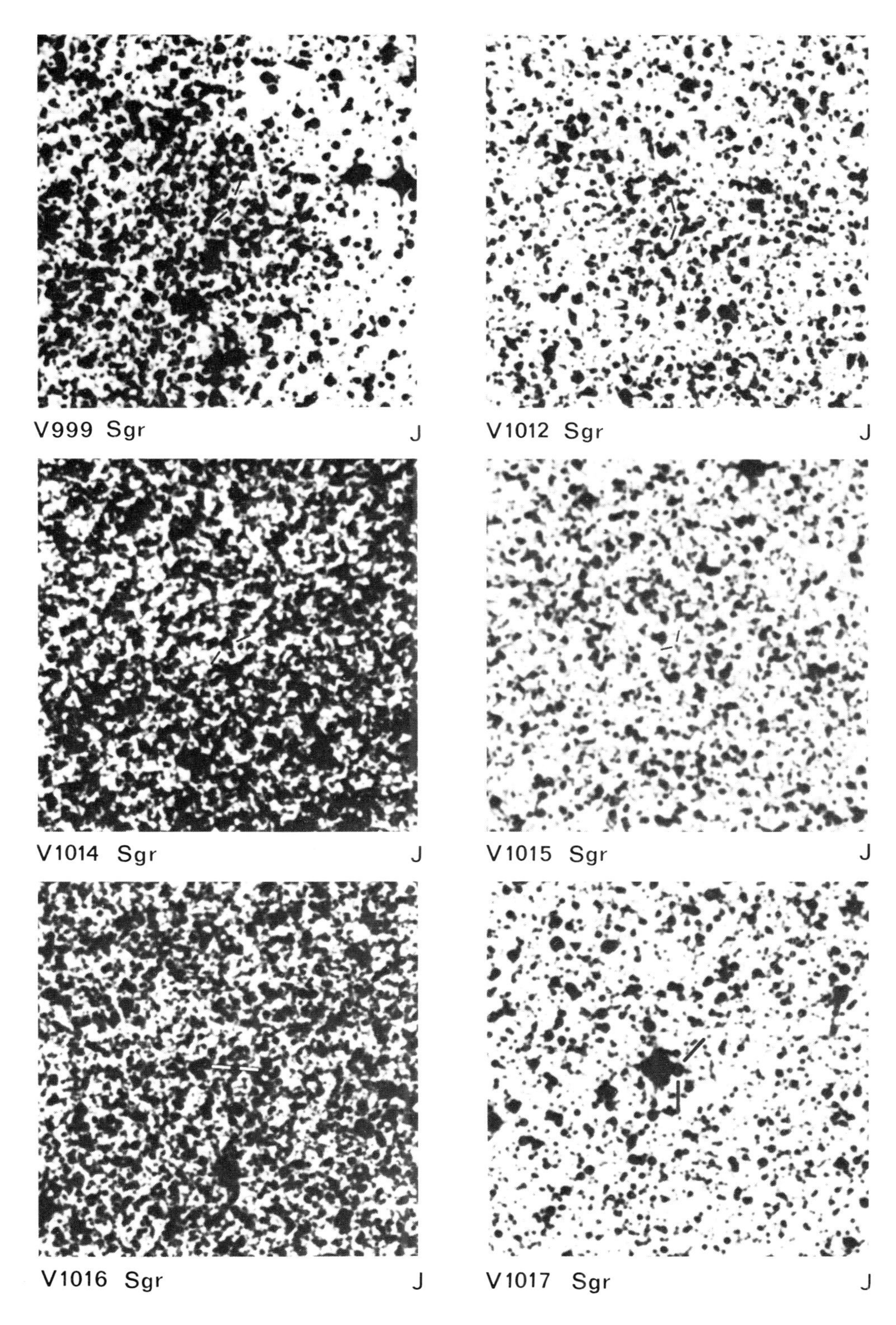
V999 Sgr J
V1012 Sgr J
V1014 Sgr J
V1015 Sgr J
V1016 Sgr J
V1017 Sgr J

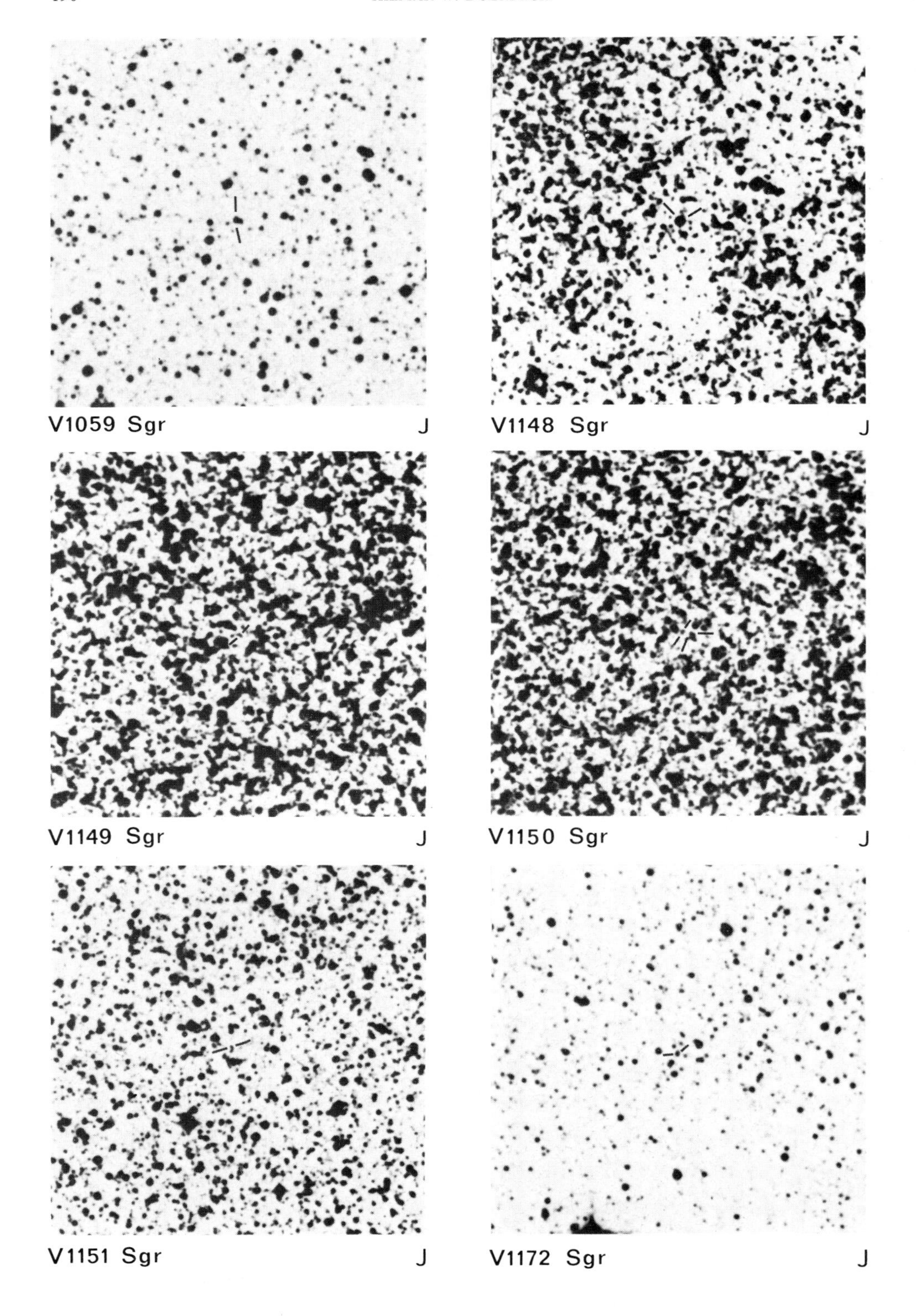
V1059 Sgr
J
V1148 Sgr
J
V1149 Sgr
J
V1150 Sgr
J
V1151 Sgr
J
V1172 Sgr
J

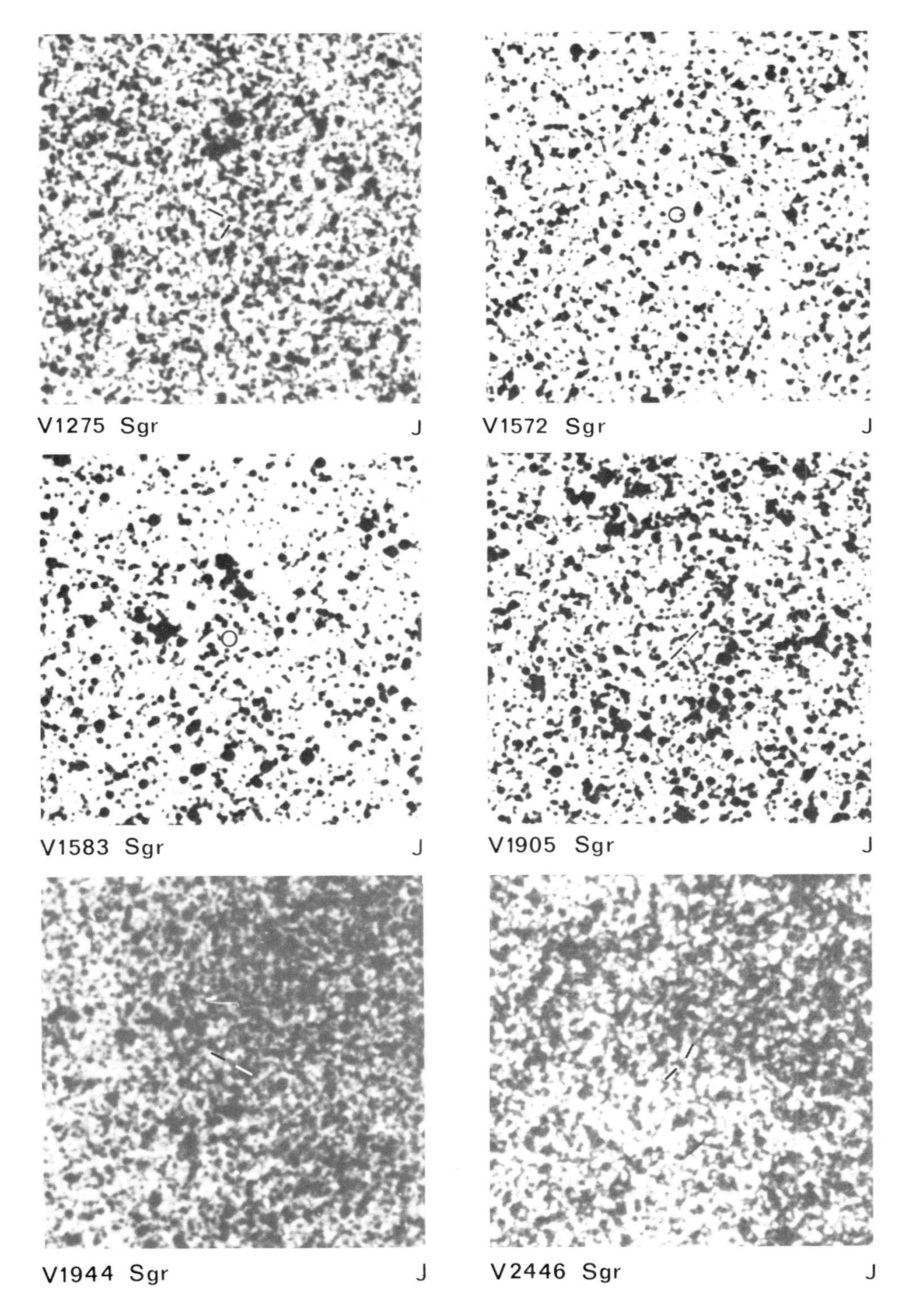
V1275 Sgr J
V1572 Sgr J
V1583 Sgr J
V1905 Sgr J
V1944 Sgr J
V2446 Sgr J

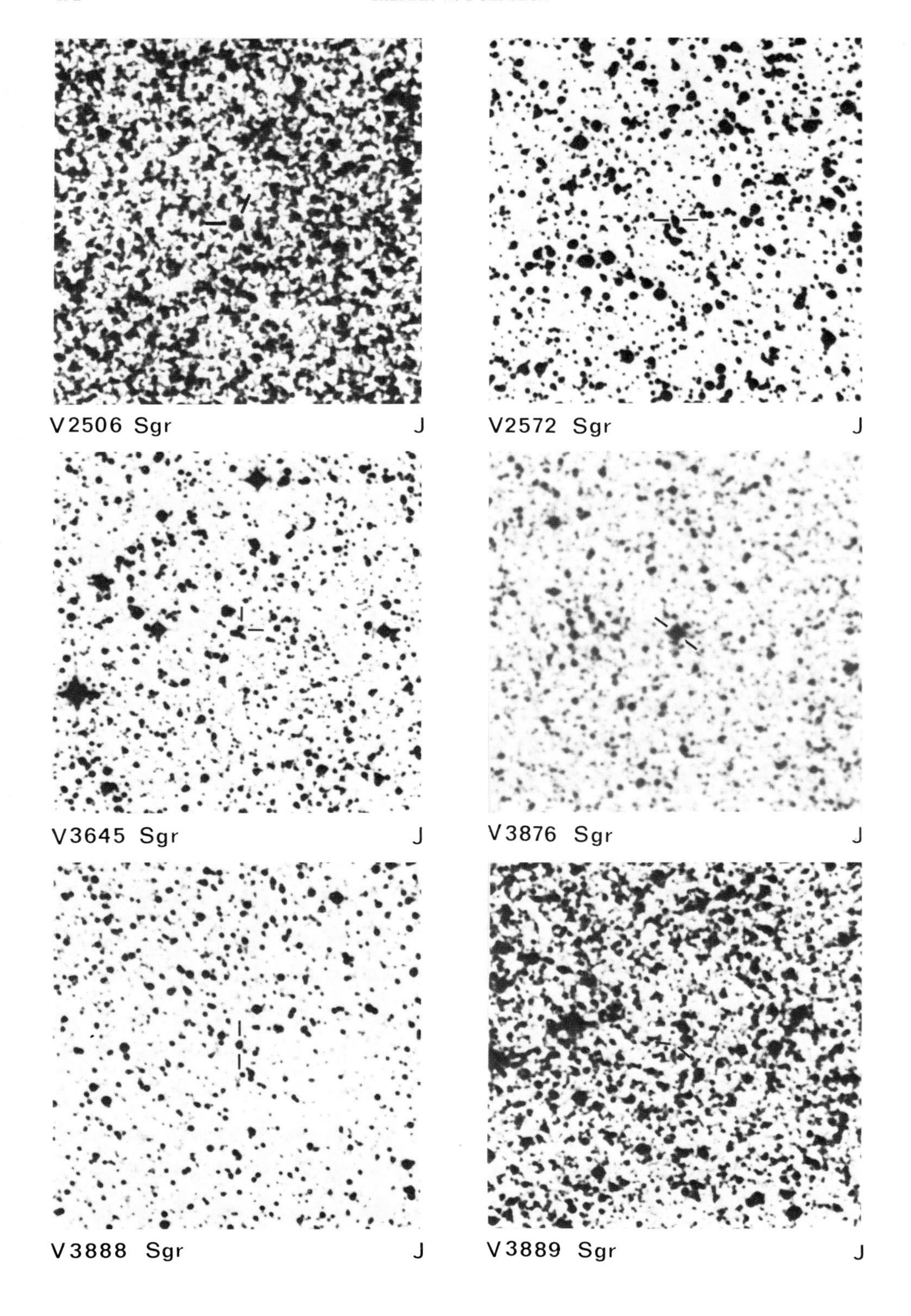
V2506 Sgr
J
V2572 Sgr
J
V3645 Sgr
J
V3876 Sgr
J
V3888 Sgr
J
V3889 Sgr
J

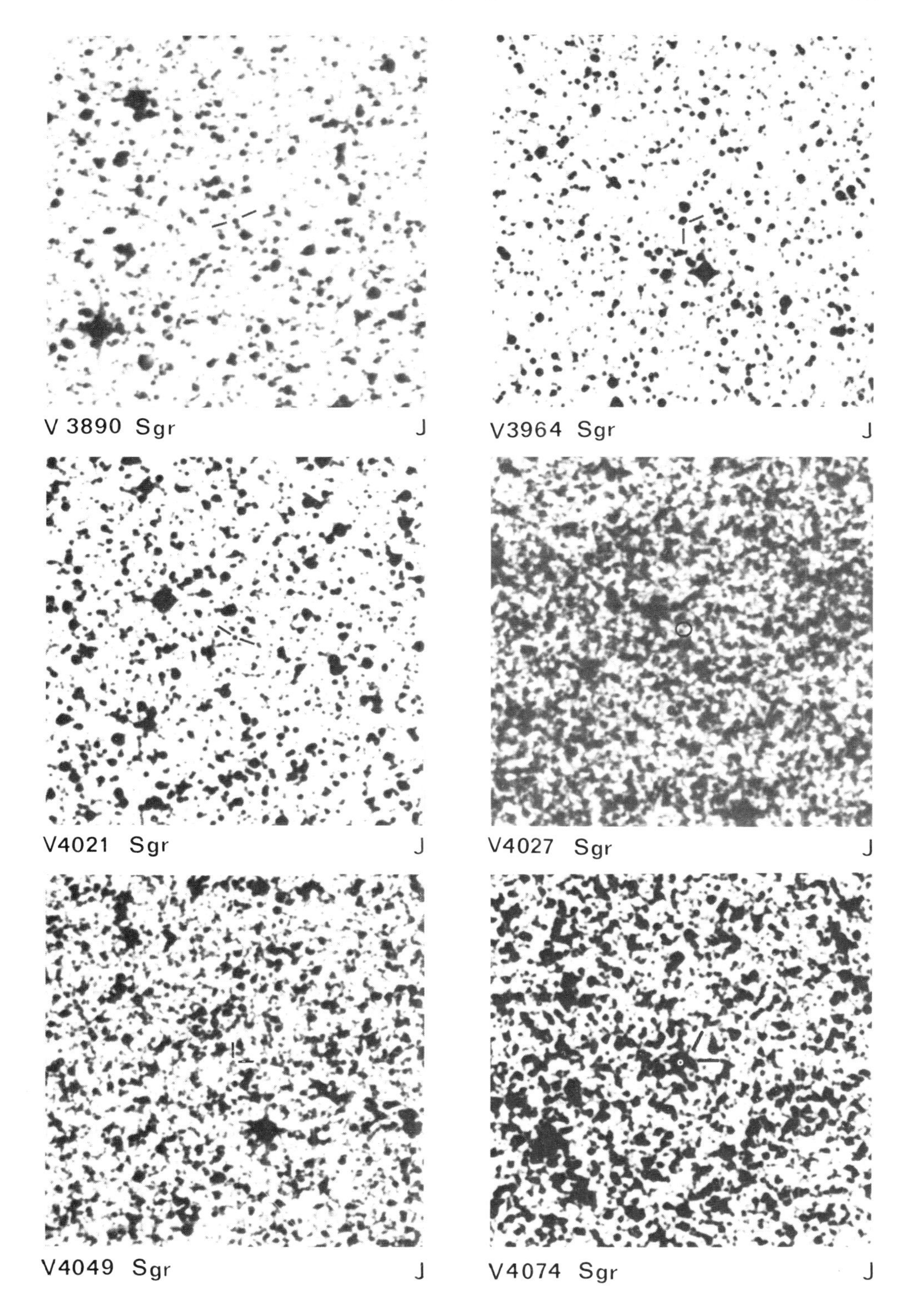
V 3890 Sgr J
V3964 Sgr J
V4021 Sgr J
V4027 Sgr J
V4049 Sgr J
V4074 Sgr J

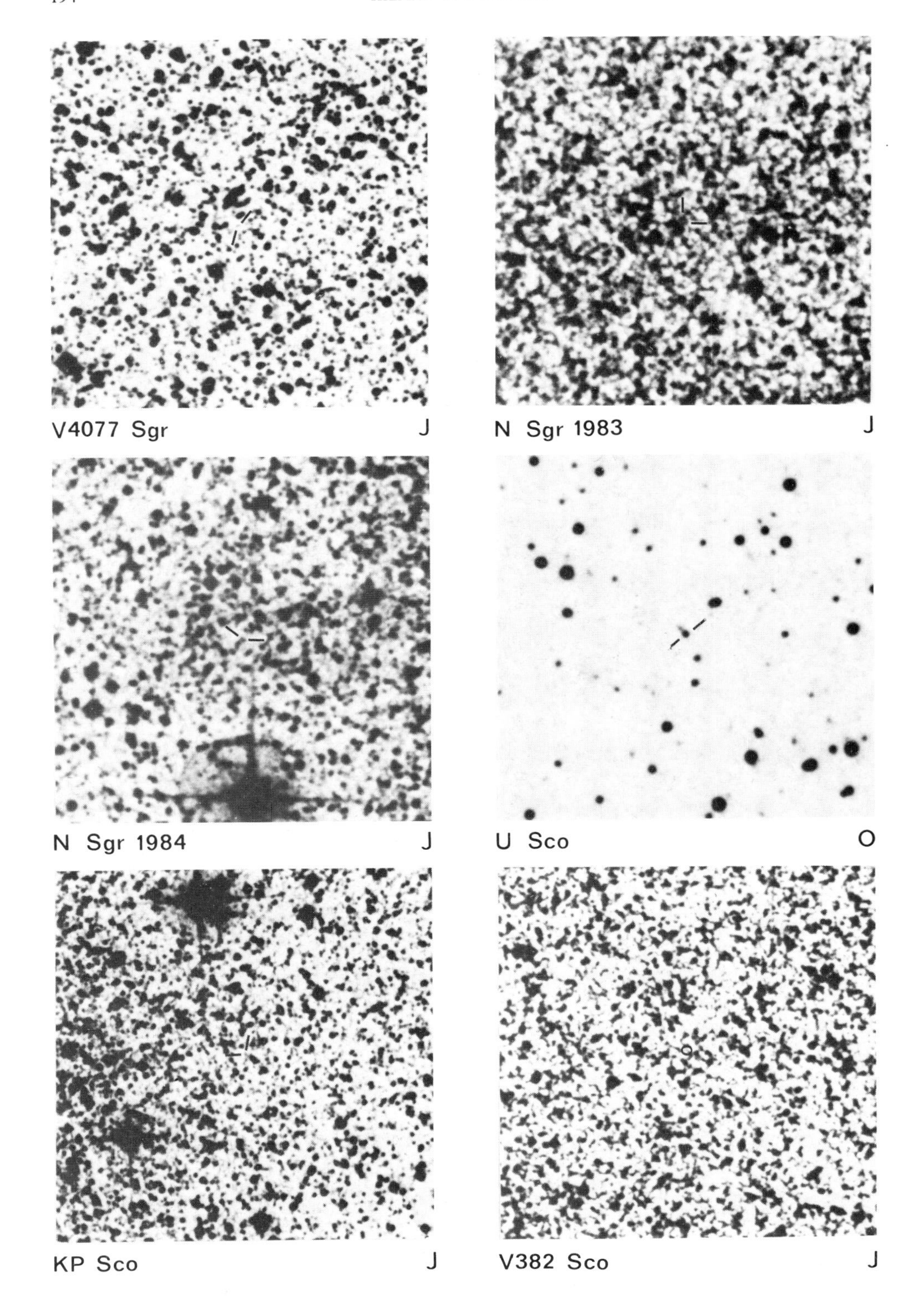
V4077 Sgr
J
N Sgr 1983
J
N Sgr 1984
J
U Sco
O
KP Sco
J
V382 Sco
J

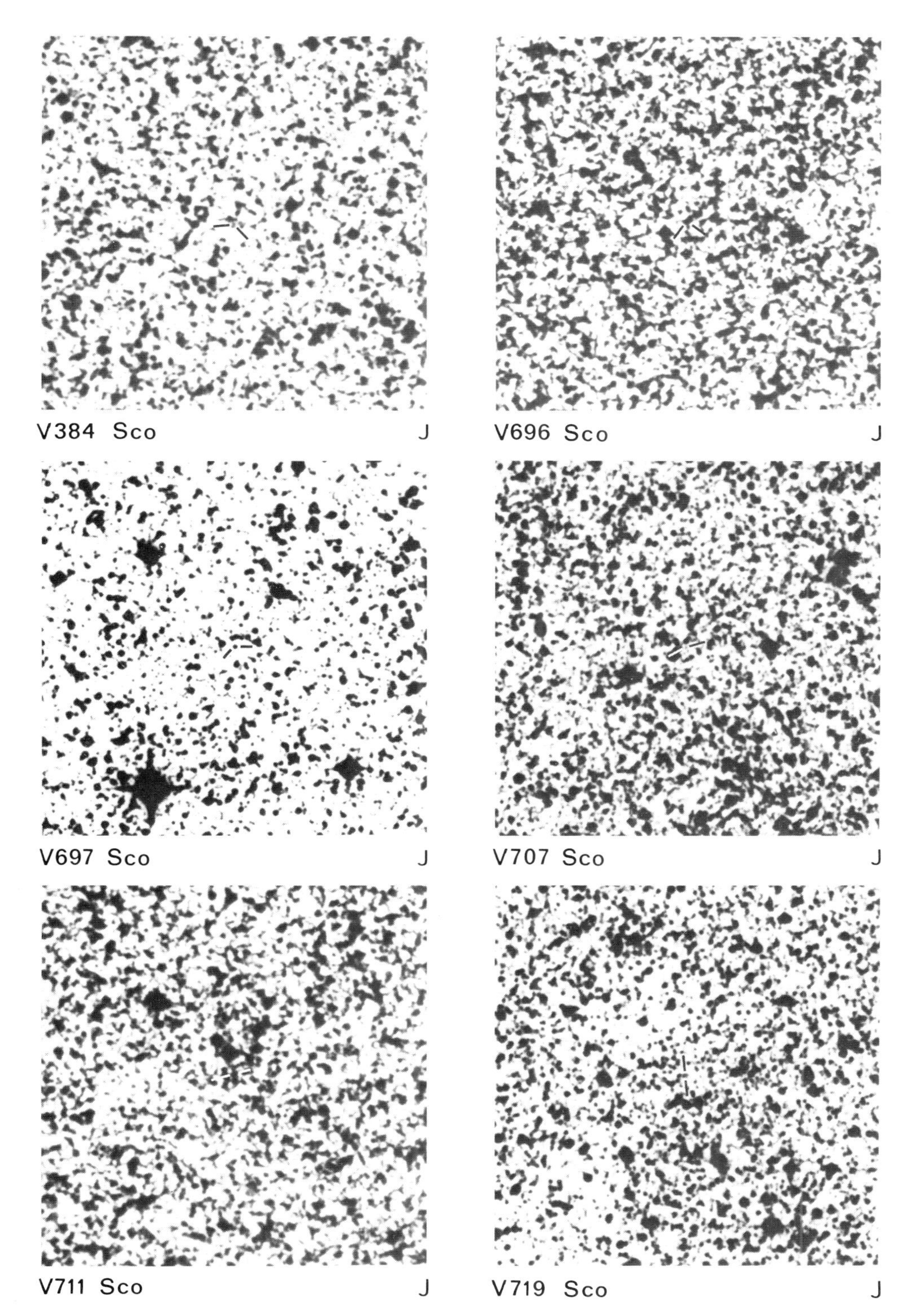

V384 Sco J

V696 Sco J

V697 Sco J

V707 Sco J

V711 Sco J

V719 Sco J

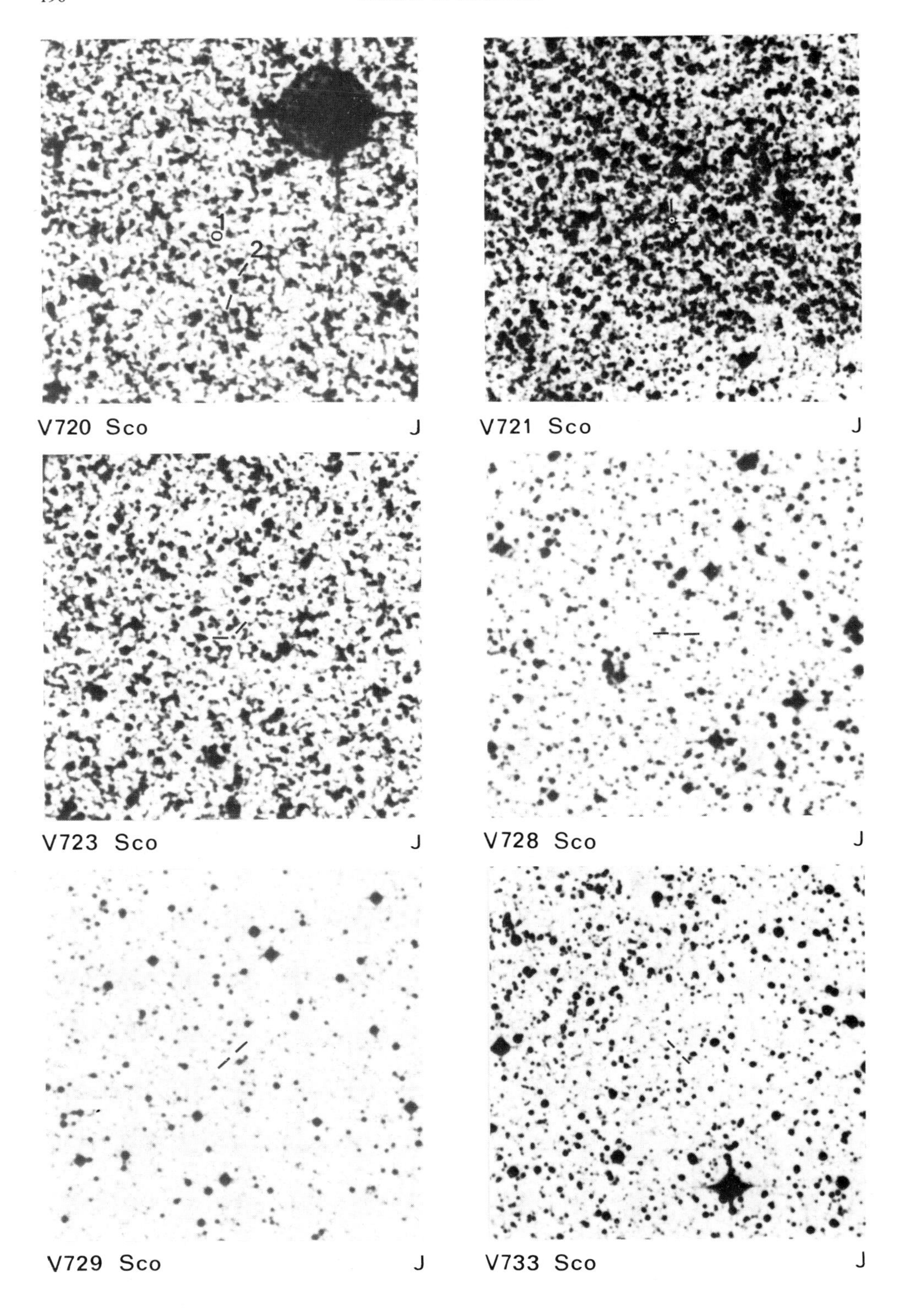
1
2
V720 Sco
J
V721 Sco
J
V723 Sco
J
V728 Sco
J
V729 Sco
J
V733 Sco
J

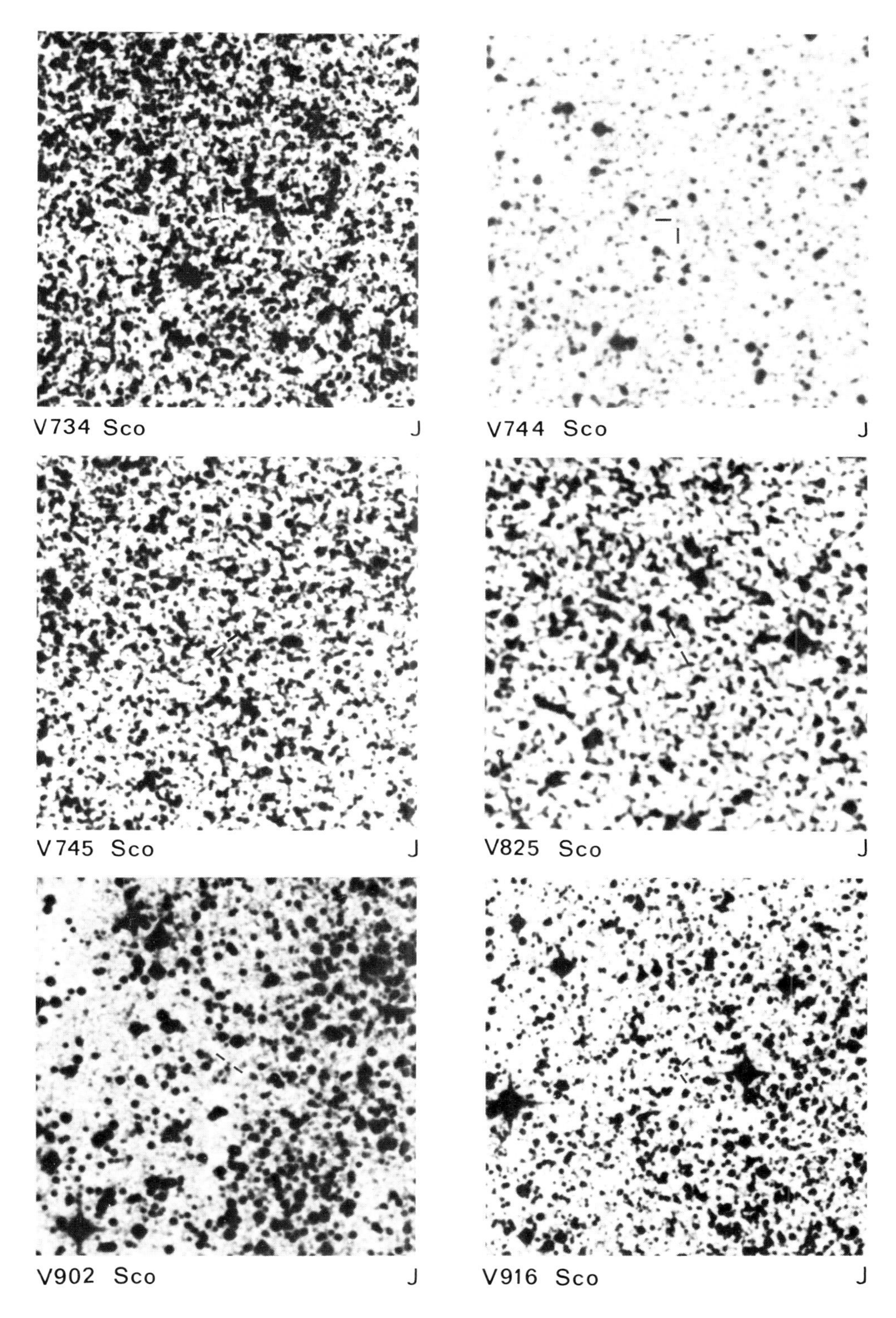

V734 Sco J

V744 Sco J

V745 Sco J

V825 Sco J

V902 Sco J

V916 Sco J

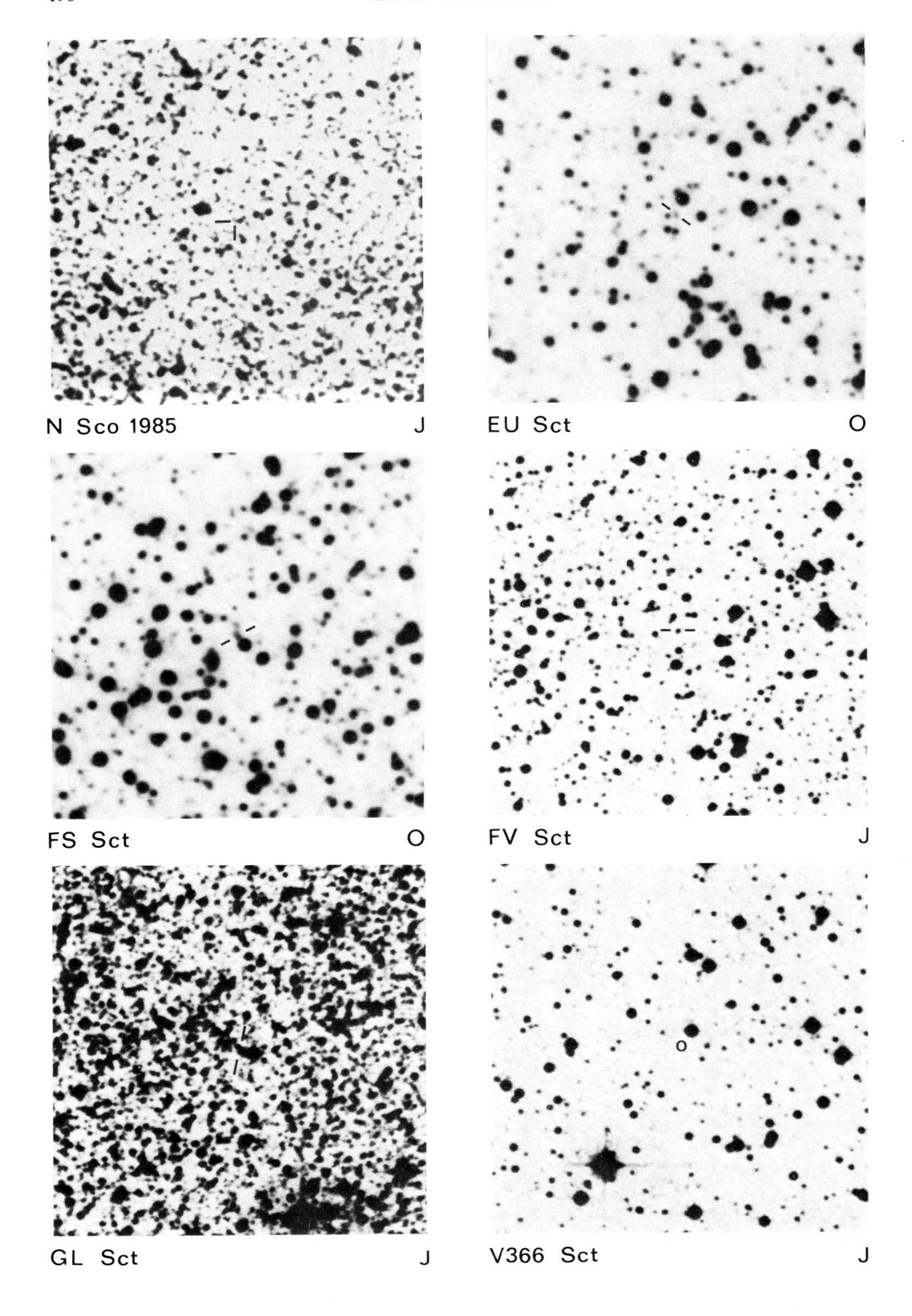
N Sco 1985 J
EU Sct O
FS Sct O
FV Sct J
GL Sct J
V366 Sct J

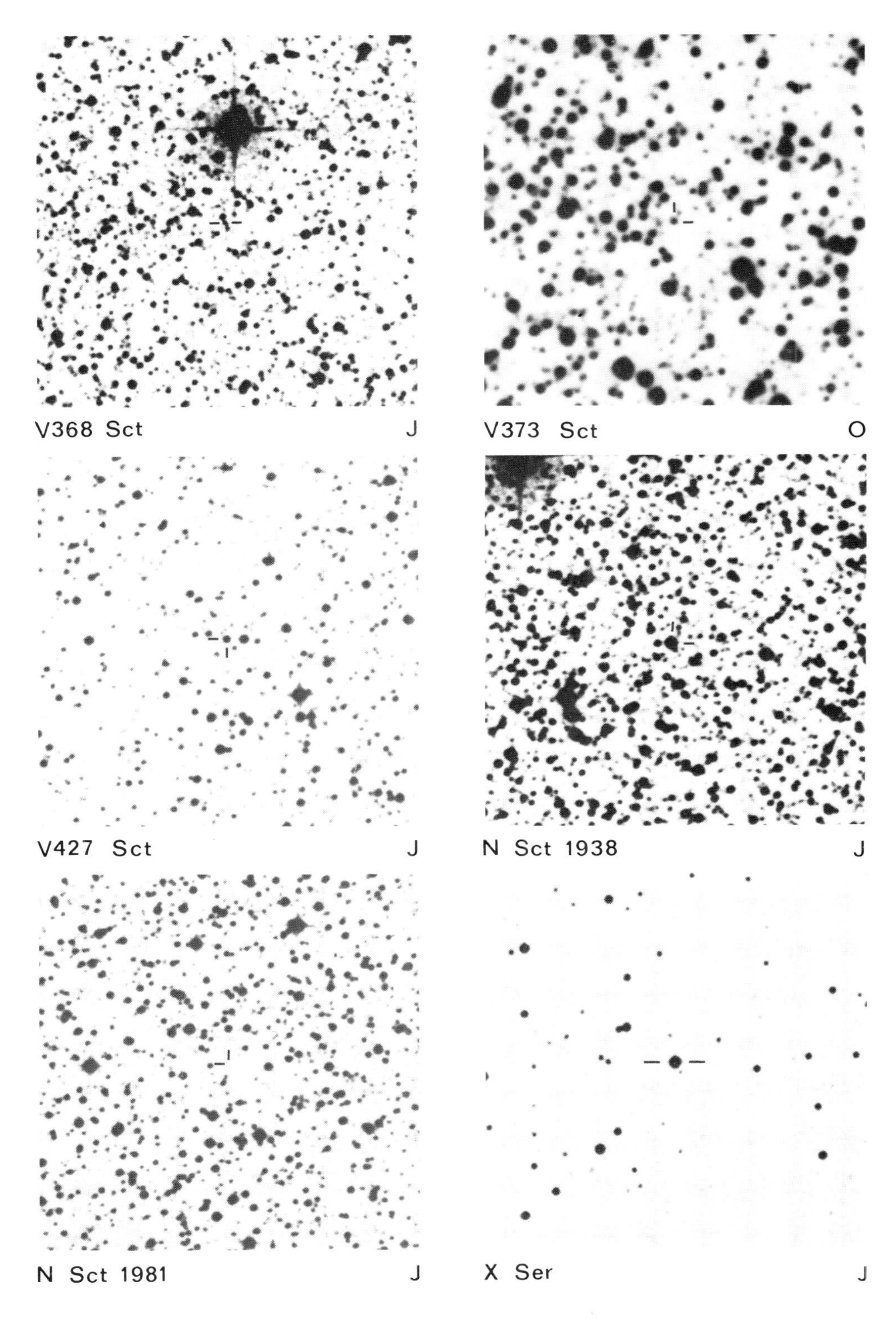
V368 Sct
J
V373 Sct
O
V427 Sct
J
N Sct 1938
J
N Sct 1981
J
X Ser
J

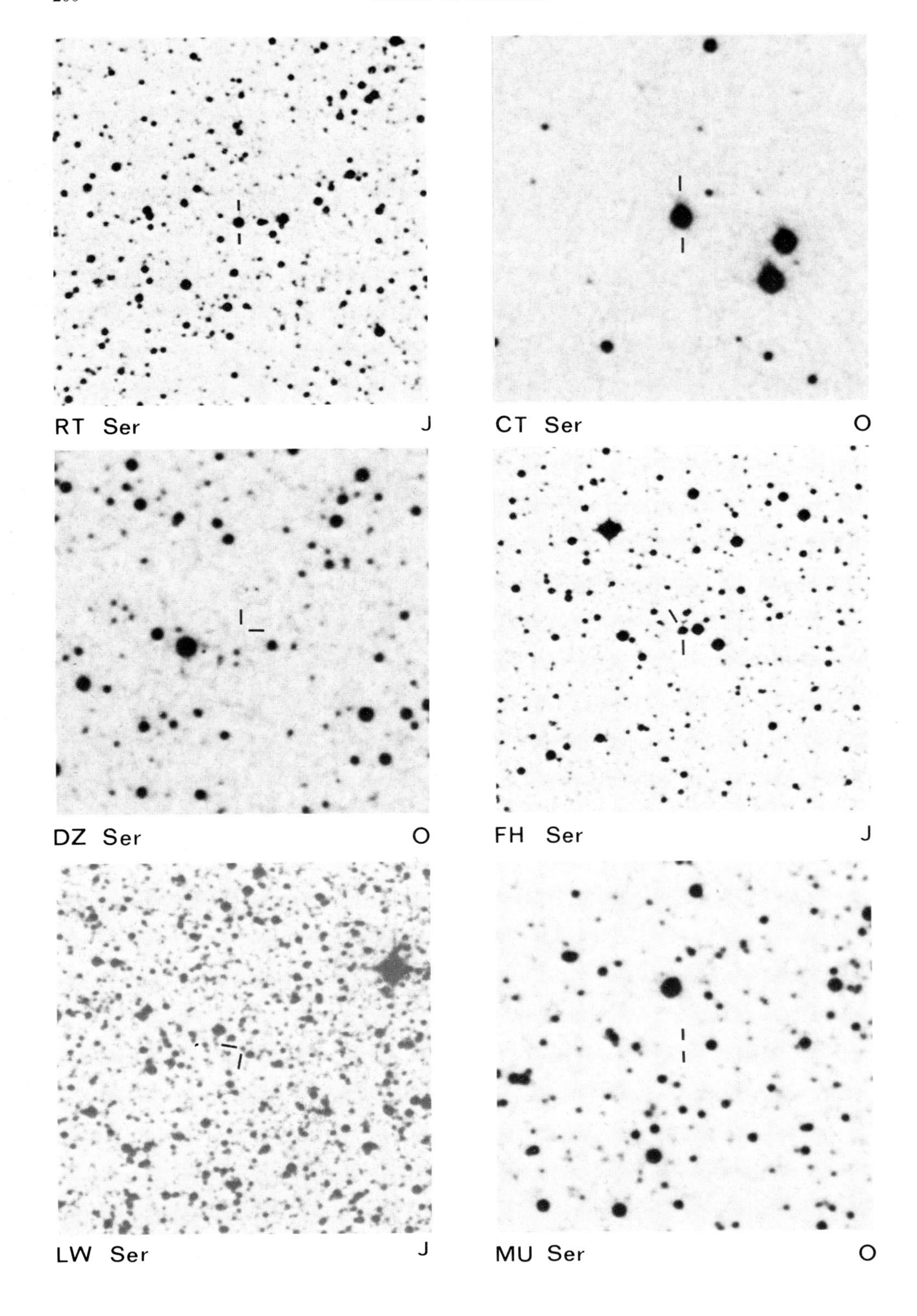
RT Ser
J
CT Ser
O
DZ Ser
O
FH Ser
J
LW Ser
J
MU Ser
O

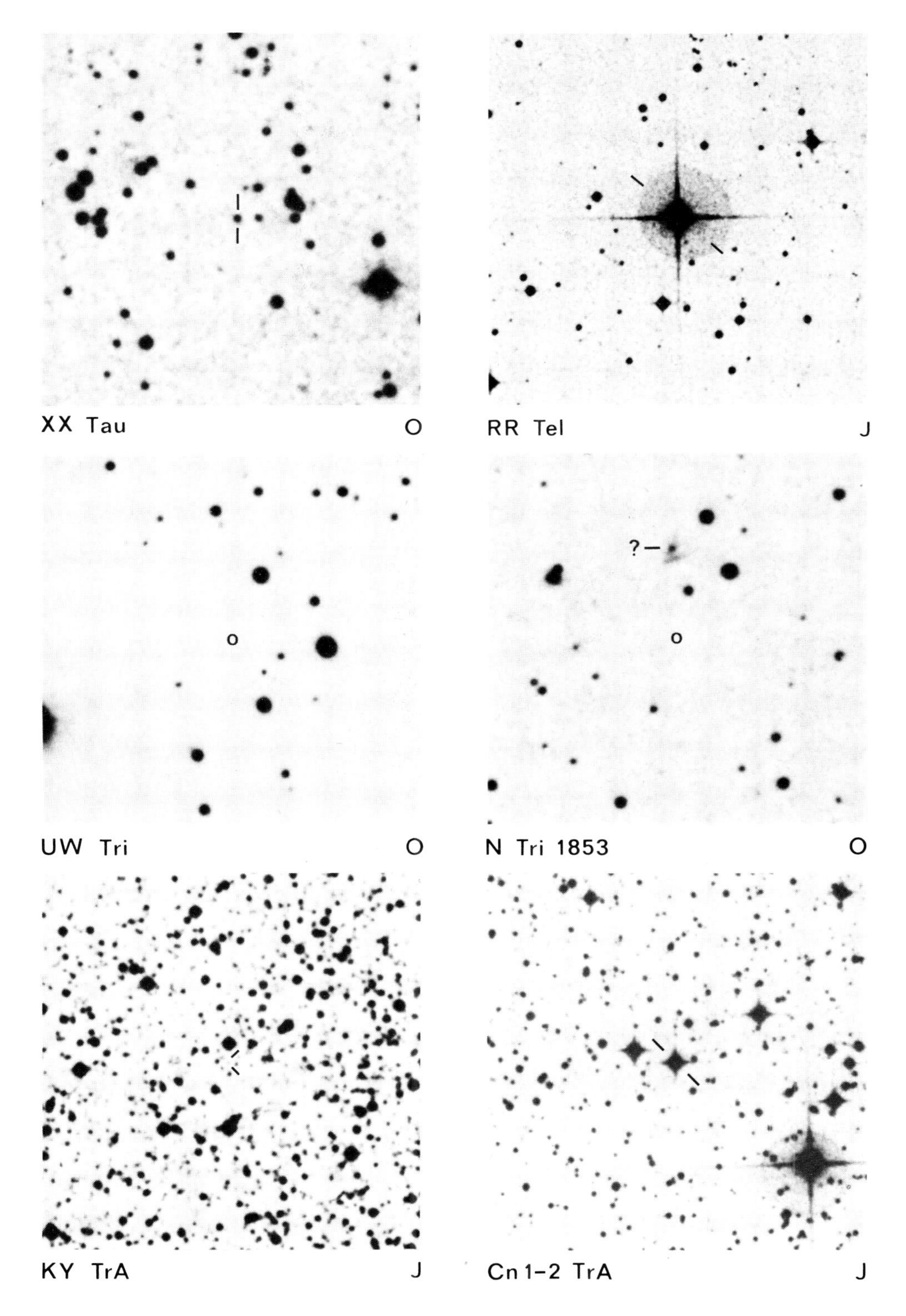

XX Tau O

RR Tel J

UW Tri O

N Tri 1853 O

KY TrA J

Cn 1–2 TrA J

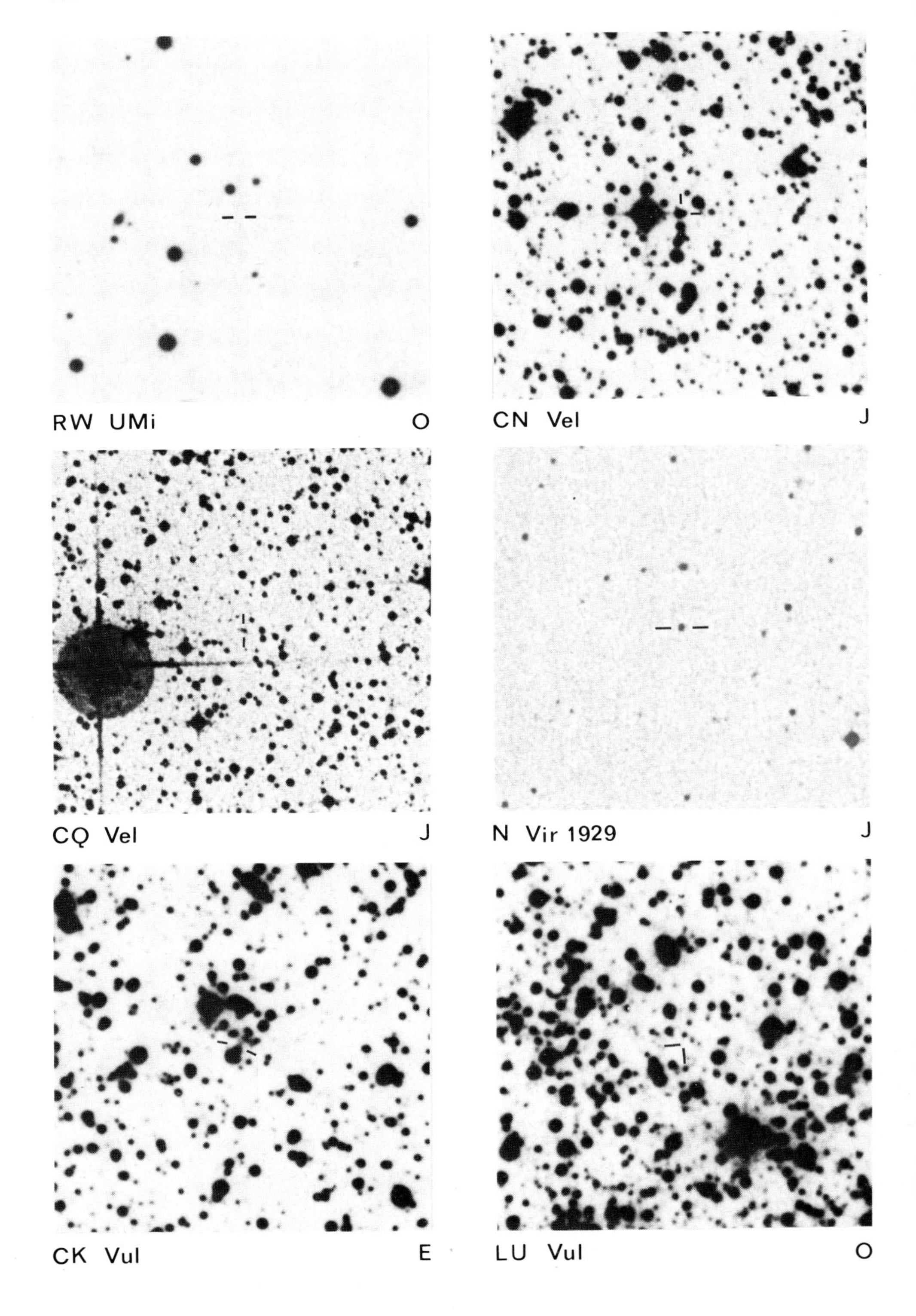
RW UMi
O
CN Vel
J
CQ Vel
J
N Vir 1929
J
CK Vul
E
LU Vul
O

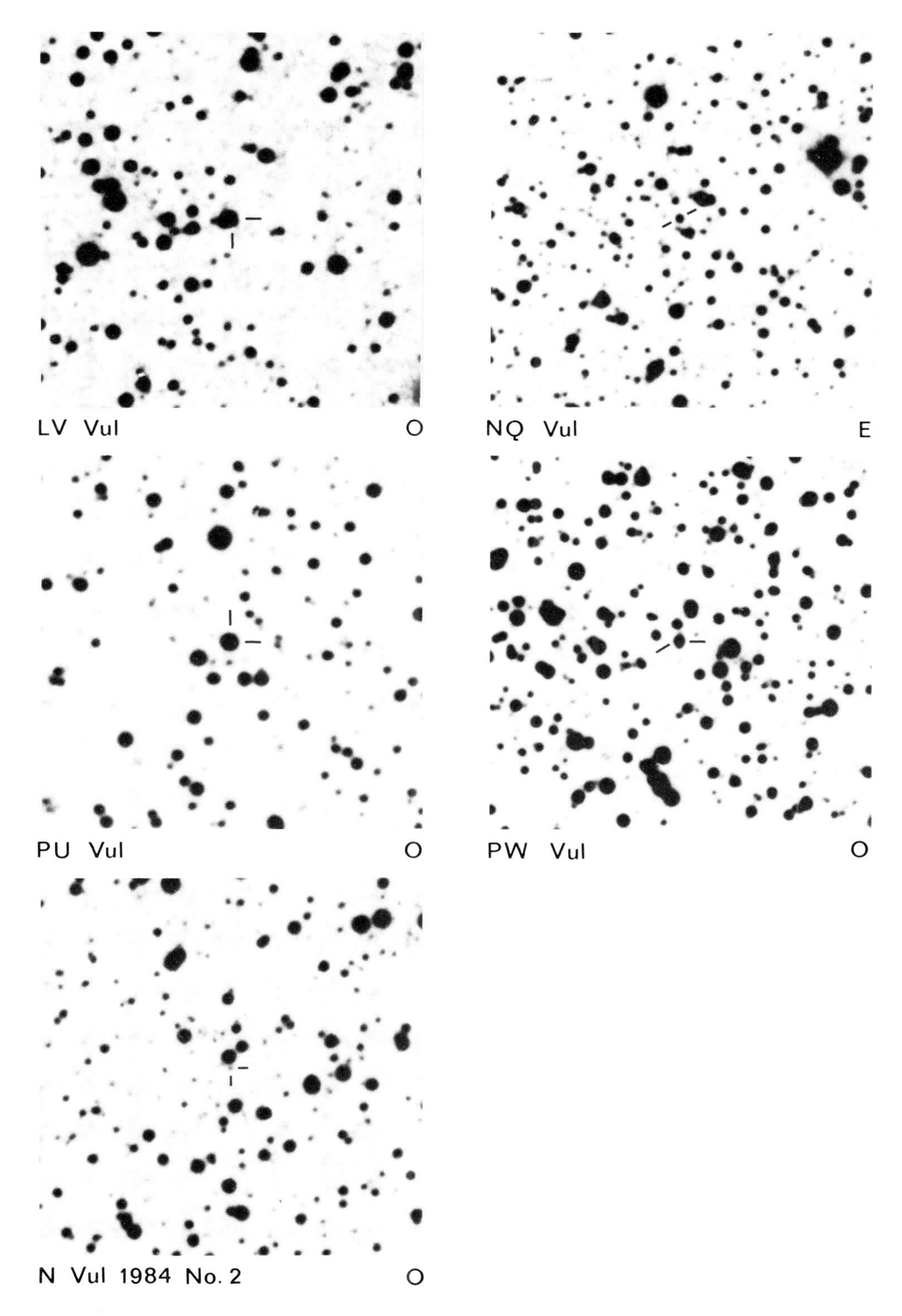
LV Vul
O
NQ Vul
E
PU Vul
O
PW Vul
O
N Vul 1984 No. 2
O

APPENDIX

to the Atlas

Field maps are given of those objects which could not be identified and for which not even candidates could be marked.

The appendix also contains finding charts for the two most recent novae, which were identified after the Atlas was completed.

Field sizes, orientations, other information and copyrights are the same as in the Atlas.

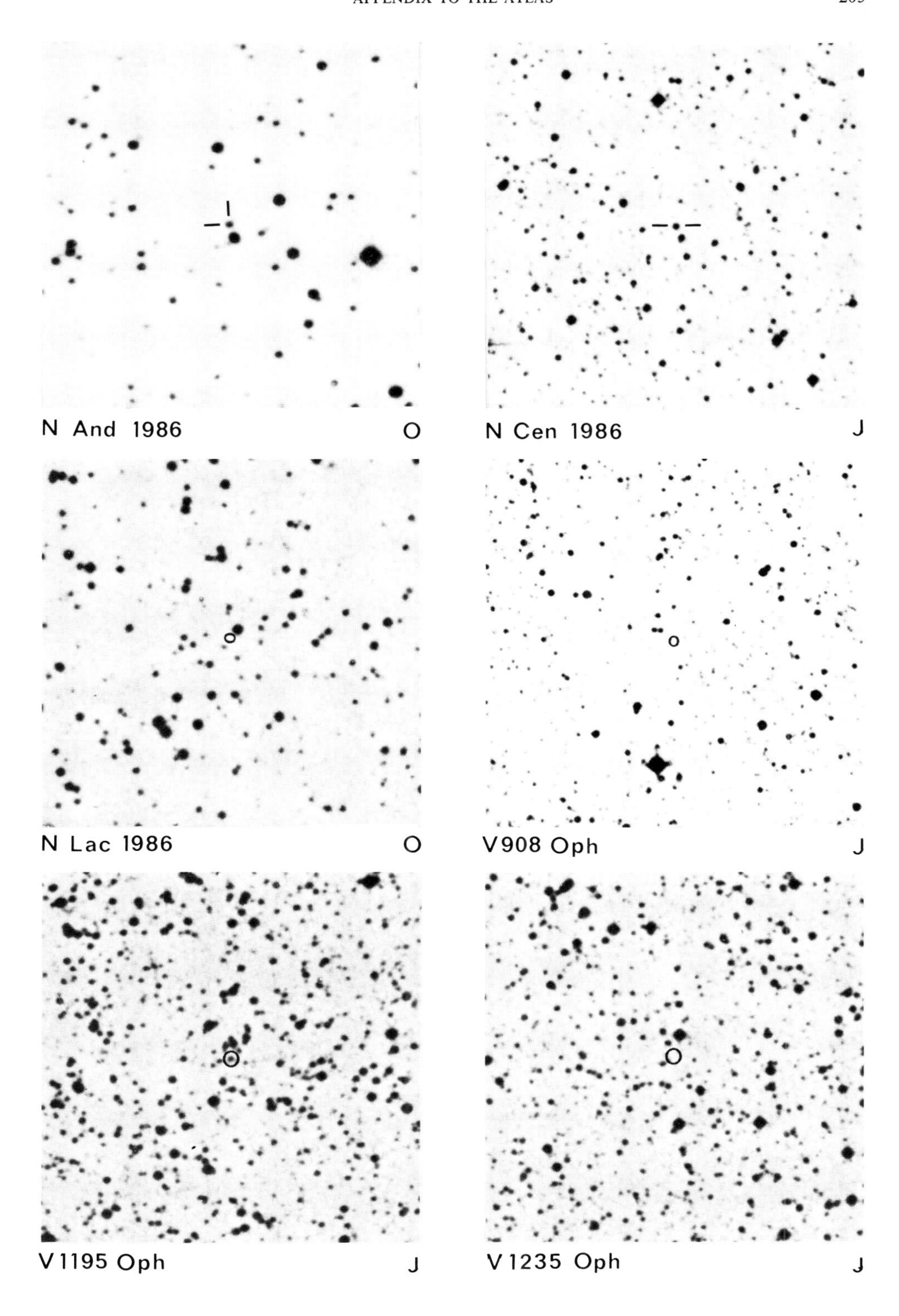
N And 1986
O
N Cen 1986
J
N Lac 1986
O
V908 Oph
J
V1195 Oph
J
V1235 Oph
J

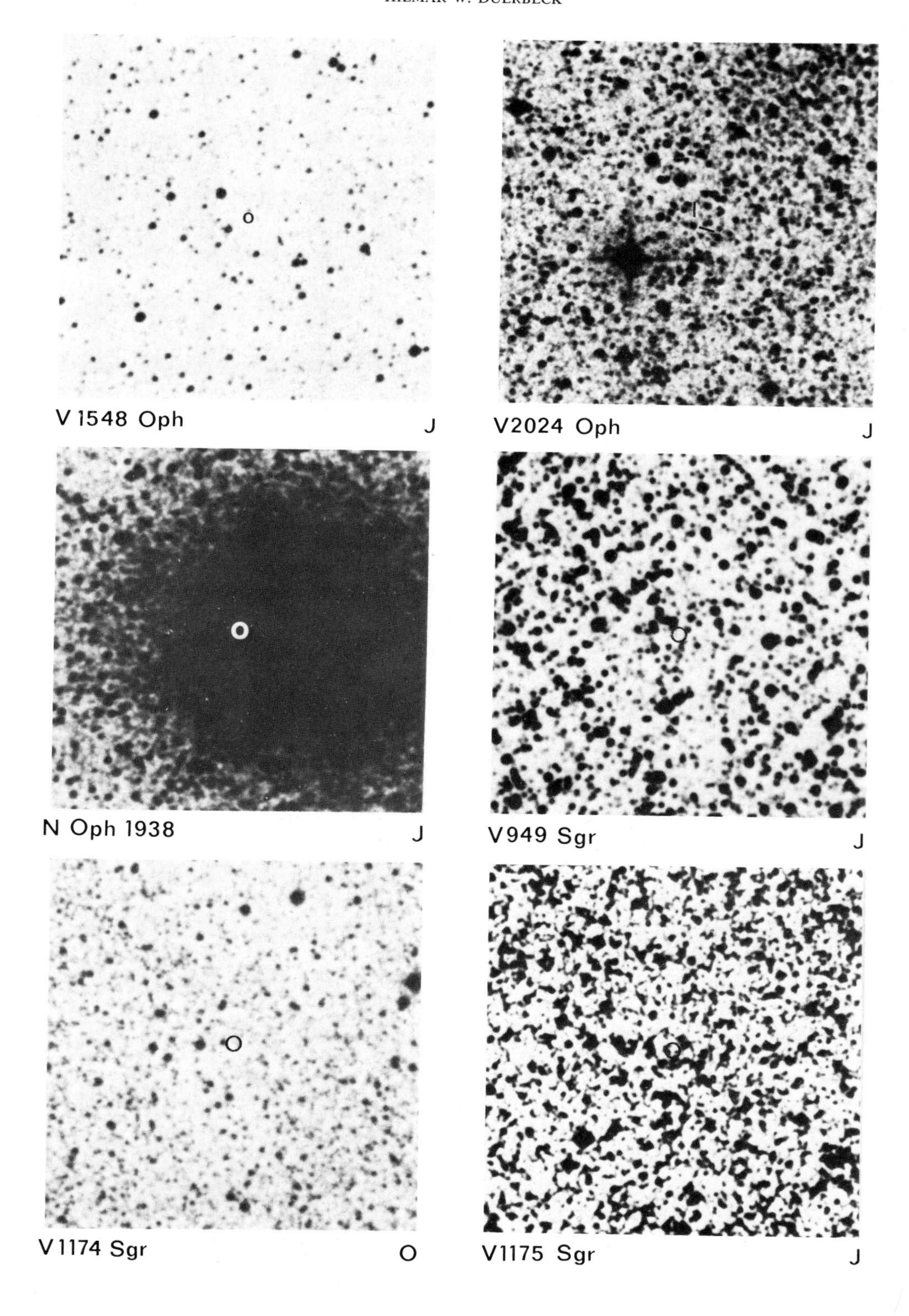
V 1548 Oph
J
V2024 Oph
J
N Oph 1938
J
V949 Sgr
J
V1174 Sgr
O
V1175 Sgr
J

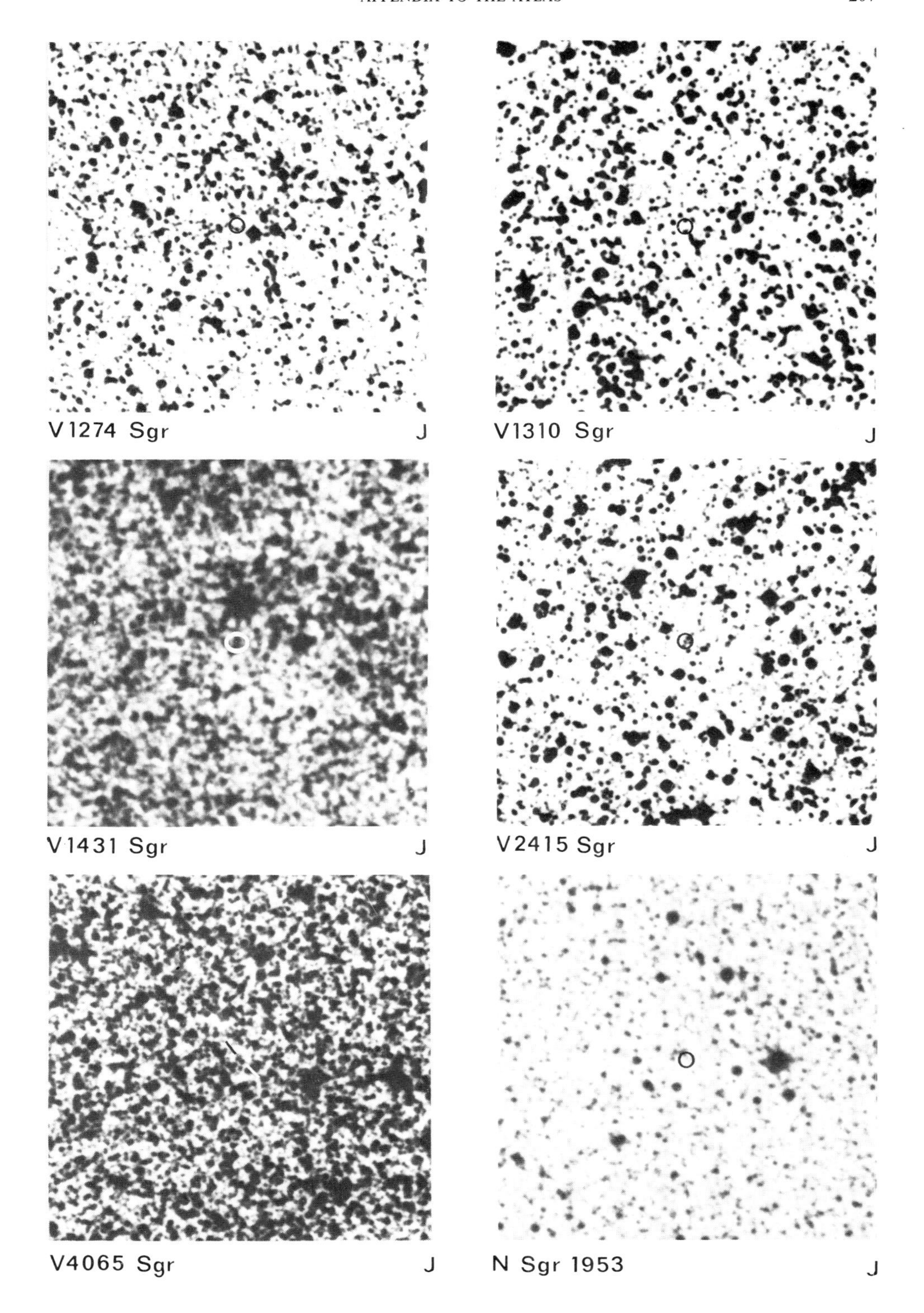

V1274 Sgr J

V1310 Sgr J

V1431 Sgr J

V2415 Sgr J

V4065 Sgr J

N Sgr 1953 J

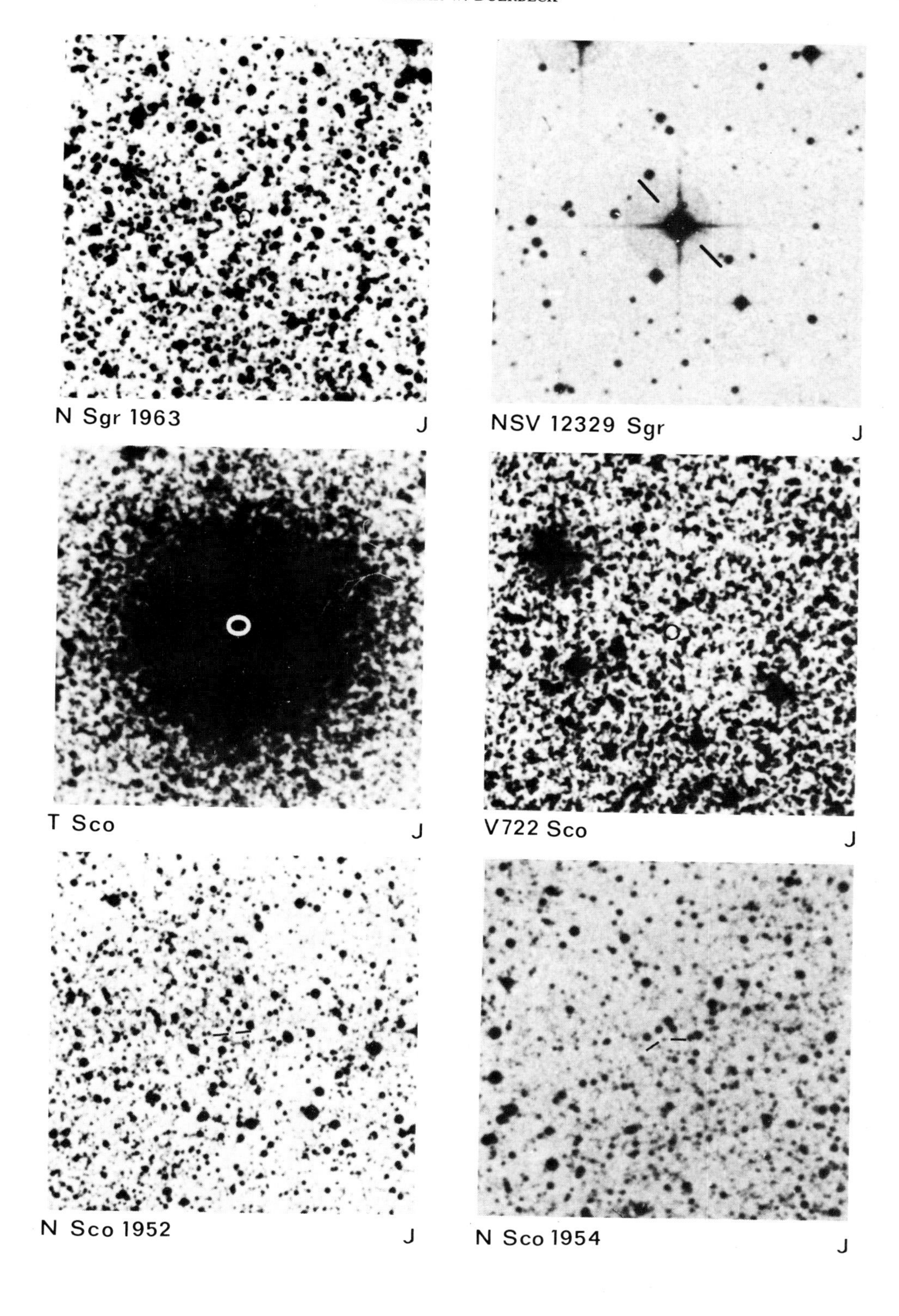
N Sgr 1963
J
NSV 12329 Sgr
J
T Sco
J
V722 Sco
J
N Sco 1952
J
N Sco 1954
J

INDEX OF NOVAE

The page numbers of the catalogue entries are given in bold letters, the page numbers of the finding charts in italics.